Discrete-continuum Coupling Method to Simulate Highly Dynamic Multi-scale Problems

Discrete Element Model and Simulation of Continuous Materials Behavior Set

coordinated by
Ivan Iordanoff

Volume 2

Discrete-continuum Coupling Method to Simulate Highly Dynamic Multi-scale Problems

Simulation of Laser-induced Damage in Silica Glass

Mohamed Jebahi
Frédéric Dau
Jean-Luc Charles
Ivan Iordanoff

iSTE

WILEY

First published 2015 in Great Britain and the United States by ISTE Ltd and John Wiley & Sons, Inc.

ISTE Ltd
27-37 St George's Road
London SW19 4EU
UK

www.iste.co.uk

John Wiley & Sons, Inc.
111 River Street
Hoboken, NJ 07030
USA

www.wiley.com

Library of Congress Control Number: 2015949734

British Library Cataloguing-in-Publication Data
A CIP record for this book is available from the British Library
ISBN 978-1-84821-771-3

Contents

List of Figures

Chapter 1

Chapter 2

Chapter 3

Chapter 4

Chapter 5

Chapter 6

List of Tables

Preface

Smart materials, added value manufacturing and factories for the future are key technological subjects for the future product developments and innovation. One of the key challenges is to play with the microstructure of the material to not only improve its properties but also to find new properties. Another key challenge is to define micro- or nanocomposites in order to mix physical properties. This allows enlarging the field of possible innovative material design. The other key challenge is to define new manufacturing processes to realize these materials and new factory organization to produce the commercial product. From the material to the product, the numerical design tools must follow all these evolutions from the nanoscopic scale to the macroscopic scale (simulation and optimization of the factory). If we analyze the great amount of numerical tool development in the world, we find a great amount of development at the nanoscopic to the microscopic scales, typically linked to *ab initio* calculations and molecular dynamics. We also find a great amount of numerical approaches used at the millimeter to the meter scales. The most famous one in the field of engineering is the finite element method. However, there is a numerical death valley to pass through, from micrometers to several centimeters. This scale corresponds to the need for taking into account discontinuity or microstructures in the material behavior at the sample scale or component scale (several centimeters). Since the 2000s, some attempts have been carried out to apply the discrete element method (DEM) for simulation of continuous materials. This method has been developed historically for true granular materials, such as sand, civil engineering grains and pharmaceutical powders. Some recent developments give new and simple tools to simulate quantitatively continuous materials and to pass from microscopic interactions at the material scale to the classical macroscopic properties at the component scale (stress and strain, thermal conductivity, cracks, damages, electrical resistivity, etc.).

In this set of books on discrete element model and simulation of continuous materials, we propose to present and explain the main advances in this field since 2010. The first book explained in a clear and simple manner the numerical way to build a DEM simulation that gives the right (same) macroscopic material properties, e.g. Young's modulus, Poisson's ratio, thermal conductivity, etc. Then, it showed how this numerical tool offers a new and powerful method for analysis and modeling of cracks, damages and finally failure of a component. In this second book, we present the coupling (bridging) between the DEM method and continuum numerical methods, such as the constrained natural element method. This allows us to focus DEM in the parts where the microscopic properties and discontinuities lead the behavior and leave continuum calculation where the material can be considered as continuous and homogeneous. Coupling scales for highly dynamic problem has been a challenge for a long time. This book shows how to choose the coupling parameters properly to avoid spurious wave reflection and to allow the passage of all the dynamic information both from fine to coarse model and from coarse to fine model. The second part demonstrates the ability of the coupling method to simulate a highly nonlinear dynamical problem: the laser shock processing of silica glass.

A further book in this set presents the numerical code developed under the free License GPL 'GranOO': www.granoo.org. All the presented developments are implemented in a simple way on this platform. This allows scientists and engineers to test and contribute to improving the presented methods in a simple and open way.

Now, dear readers let us open this book and welcome in the DEM community for the material of future development ...

Ivan IORDANOFF
Bordeaux, France
August, 2015

Introduction

I.1. Bridging the scales in science and engineering

Over the past few decades, numerical simulation has firmly established itself as a partner to experiment with unraveling the fundamental principles behind continuous material behaviors. Starting from the 1960s, this approach received strong scientific interest which led to the development of a great number of numerical methods. These methods can be divided into continuum methods (CMs) and discrete methods (DMs). Undoubtedly, the CMs are the most commonly used to solve problems at the engineering (macroscopic) scale, at which the mechanical behavior of materials can generally be described by continuum mechanics. However, their application to investigate microscopic effects, which can have a profound impact on what happens at larger space and time scales, faces several difficulties. Although solutions that are more or less reliable have been proposed in the literature to get over these difficulties, an accurate description of numerous engineering problems remains very challenging for CMs. Some difficulties, associated with reliance of these methods on a predefined mesh and/or unsuitability in dealing with discontinuities, are still not adequately ironed out. In contrast, the DMs naturally provide solutions for most of these outstanding difficulties. These are based on discrete mechanics and do not rely on any kind of mesh. Using DMs, the studied domain is modeled by a set of discrete bodies allowing discontinuities to be naturally taken into account. Although these methods were originally developed to study naturally discrete problems, their features have been proven to be very attractive for several continuous problems involving complex microscopic effects, e.g. damage, fracture and fragmentation. Application of such methods to overcome the CM limitations is then well worth exploring. Nevertheless, the lack of theoretical framework allowing these methods to properly model continua has restricted their application on this kind of problem until very recently.

Modeling continuous problems with DMs mainly faces two significant challenges. The first challenge concerns the choice of the cohesive links between the neighboring discrete bodies and the identification of their microscopic parameters so as to ensure the expected macroscopic mechanical behavior. The second challenge concerns the construction of the discrete domain which must take into account the structural properties of the original problem domain, e.g. homogeneity and isotropy, and must ensure independence of the macroscopic mechanical behavior on the discrete bodies number. The first book of this set, *Discrete Element Method to Model 3D Continuous Materials* [JEB 15], tried to tackle these challenges and to provide a comprehensive methodology allowing for correct discrete element modeling of continuous materials. This methodology was developed for a particular discrete element method (DEM) in which a given material is modeled by a set of rigid spheres in interaction with each other by three-dimensional (3D) cohesive beam bonds. As shown in [JEB 15], several conditions must be satisfied to properly model continua using the proposed DEM variation. The development of this DEM variation, in addition to the ever-increasing power and affordability of fast computers, has brought discrete element modeling of continuous material within reach. Nowadays, such a method presents a prominent tool for elucidating complex mechanical behaviors of continuous materials [AND 12b, JEB 15]. It was successfully applied to investigate several challenging problems that cannot be easily treated by CMs [AND 13, AND 12a, TER 13, JEB 13a, JEB 13b]. The major drawback of this method is that it is very time-consuming compared to CMs and the computation time can quickly become crippling, especially in the case of a large studied domain. However, in modern material science and engineering, real materials usually exhibit phenomena requiring multi-scale analysis. These phenomena require on one scale a very accurate and computationally expensive description to capture the complex effects at this scale and on another scale a coarser description is sufficient and, in fact, necessary to avoid prohibitively large computation. Therefore, in a view of expanding the scope of proposed DEM and alleviating its limitations, it would be beneficial to couple this approach with a CM, such that the computation effort can be distributed as needed.

In many mechanical problems, the notion of multi-scale modeling arises quite naturally. Indeed, most of the material behaviors at the macroscopic scale, which is the scale of interest for engineering applications, are determined by microscopic interactions between atoms. This is why such a notion has become a special area of interest for many scientists. Consequently, several multi-scale coupling methods have been developed over the last three decades. In a pioneer work, Ben Dhia [BEN 01, BEN 05, BEN 98] developed the Arlequin approach as a general framework which allows the intermixing of various mechanical models for structural analysis and computation. Abraham *et al.* [ABR 98, BRO 99] developed a methodology that couples the tight-bending quantum mechanics with molecular dynamics (MD) such that the two Hamiltonians are averaged in a bridging region. A damping was used in this region to reduce spurious reflections at the interface between the two models. Nevertheless, the choice of the damping coefficient remains difficult. Smirnova *et al.* [SMI 99] proposed a combined MD and finite element method (FEM) model with a transition zone in which the FEM nodes coincide with the positions of the particles in the MD region. The particles in the transition zone interact with the MD region via the interaction potential. At the same time, they experience the nodal forces due to the FEM grid. Belytschko and Xiao [BEL 03, XIA 04] developed a coupling method between the molecular dynamics and continuum mechanics models based on the bridging domain technique. In this method, the two models are overlaid at the interface and constrained with a Lagrange multiplier model in the bridging region. Fish *et al.* [FIS 07] formulated an atomistic-continuum coupling method based on a blend of the continuum stress and the atomistic force. In terms of equations, this method is very similar to the Arlequin approach [BEN 01, BEN 05, BEN 98]. In an interesting work, Chamoin *et al.* [CHA 10] analyzed the main spurious effects in the atomic-to-continuum coupling approaches and they proposed a corrective method based on the computation and injection of dead forces in the Arlequin formulation to offset these effects. Aubertin *et al.* [AUB 10] applied the Arlequin approach to couple the extended finite element method (X-FEM) with MD to study dynamic crack propagation. Bauman *et al.* [BAU 09] developed a 3D multi-scale method, based on the Arlequin approach, between highly heterogeneous particle models and nonlinear elastic continuum models. For more details, several papers reviewing these methods can be found in the literature [LU 05, XU 09, JEB 14, CUR 03]. Based on these papers, three approaches can mainly be used to couple DEM with CMs: the hierarchical, concurrent and hybrid hierarchical-concurrent coupling approach.

The hierarchical coupling approach, also called sequential, serial, implicit or message passing, is the most widely used and computationally the most efficient. This approach aims to piece together a hierarchy of numerical methods in which the coarse-scale model uses information obtained by the more detailed fine-scale model. The homogenization methods for multi-phase media are typical examples of the hierarchical coupling approach. The response of a representative volume element at a fine scale is first computed, and from this, a stress-strain law is extracted to describe the mechanical behavior of the homogenized material at coarser scale. The hierarchical approach is generally well suited for problems in which the different analysis scales are decoupled or weakly coupled. In other words, it can be used when the large-scale variations appear homogeneous and quasi-static from the fine scale point of view.

The concurrent coupling approach, also called parallel or explicit, consists of linking numerical models of different scales together in a single combined model, such that the fine-scale model communicates directly and instantly with the coarse-scale model through some coupling procedure. Both compatibility and momentum balance are enforced across the interface between the coupled models. This type of coupling approach is well suited to study multi-scale problems in which the behavior at each scale depends strongly on what happens at the other scale. A variation of the concurrent approach, generally referred to as semi-concurrent, is that in which the coupled models run together and communicate instantly with each other but are not intimately coupled. Compatibility and momentum balance are only satisfied approximately. The advantages of this approach lie in the fact that the coupled models can be computed by separate software. The FE^2 multi-scale approach of Feyel and Chaboche [FEY 00] is an example of this variation.

In some multi-scale problems, the involved scales can be weakly coupled at the beginning of the computation up to a certain response limit, and subsequently become highly dependent. Therefore, it would be computationally beneficial to combine the above two coupling approaches to study such problems. This has led to the development of the hybrid hierarchical-concurrent (or hierarchical-semi-concurrent) coupling approach. The hierarchical approach is used as long as the requested fine-scale information is available. When this information is no longer accessible, due to strain localization, for example, the concurrent (or semi-concurrent) approach is invoked. If the fine-scale response is history dependent, it is necessary to reconstruct at least an approximate history. An example of this approach is the adaptive multi-scale approach developed by Akbari *et al.* [AKB 12] to study the quasi-brittle crack propagation in metals. In this approach, the FE^2 technique is used in the safe regions of the studied domain. When strain localization appears, the concurrent coupling approach is used to solve the problem exactly at the material heterogeneities scale. Compared to the first two coupling approaches, the hybrid hierarchical-concurrent (hierarchical-semi-concurrent) coupling approach is relatively recent, and is, at present, the subject of several studies [AKB 12, NUG 07].

The choice of one of the above multi-scale approaches to couple DEM and CMs depends on the type and nature of the mechanical problems to be studied. This point will be discussed in the next section.

I.2. Scope and objective

The ever-accelerating progress in applied science and engineering has given rise to numerous interesting problems that require multi-scale modeling to accurately handle the relevant phenomena, while reducing the computation time. Of particular interest are the fast dynamic problems which generally involve strongly dependent multi-scale effects. Due to their complexity, these problems so far present a central issue for traditional numerical methods. In contrast, the DEM variation proposed [JEB 15], which is well adapted for highly dynamic analysis, can give answers to several outstanding questions related to such problems. In order to benefit from the DEM strengths in solving such problems, this book focuses on this type of problem. A common feature of most of these problems is that the regions requiring fine-scale analysis by DEM are generally small with respect to the full studied domain. Modeling such problems with DEM-CM coupling approach can thereby considerably reduce the computation costs without affecting the solution accuracy. Therefore, the first objective of this book aims to develop a robust multi-scale discrete-continuum coupling approach between DEM and a CM, adapted for highly dynamic problems. Among the cited coupling approaches, the concurrent approach offers several potential benefits with regard to this objective. Indeed, this approach is the most suitable to model strongly dependent multi-scale phenomena which are frequently encountered in complex highly dynamic problems. This approach is then retained to couple DEM with a CM to be chosen. As already known, numerous CMs used to model material behaviors can be found in the literature [LIU 03, LUC 77, ZIE 05c, ZIE 05a, ZIE 05b, CHI 11]. Each method has its features and specificities. The choice of the most appropriate CM to be coupled with DEM is thus not straightforward. To simplify this task, the most commonly used CMs will be classified according to their advantages and drawbacks with respect to the aim of the present work. Based on this classification, the method that best meets the expectation of this work will be retained for coupling with DEM. Then, a concurrent coupling approach adapted for highly dynamic multi-scale problems will be developed between these two methods.

With the development of the discrete-continuum coupling approach, several interesting complex applications in fast dynamics become affordable. One particular application is the laser shock processing (LSP) of materials. Since its first industrial application in the 1970s, LSP has become widely used in various engineering areas to improve the near-surface mechanical properties of metals, to remove matter by cutting or drilling, to harden or texture surfaces, etc. The ever-increasing use of this process has created a need for more in-depth studies to deal with some outstanding challenges. Despite the current experimental advances in this direction, some of these challenges still remain to be solved. Therefore, numerical simulation has become essential to support the experimental studies. To numerically study an LSP test from a mechanical point of view, it is necessary to know the mechanical loading generated by the laser–matter interaction. Several models aiming to approximate this loading exist in the literature [MAI 08, KHA 05, COL 06, FRO 93]. However, the various simplifying assumptions underlying these models have made them inaccurate in the general case. Application of the developed discrete element method-constrained natural element method (DEM-CNEM) coupling approach as an inverse technique to enrich the results of these models, based on the final experimental results of an LSP test, would be an avenue worth exploring. Nevertheless, implementation of this idea requires preliminary simulations to ensure that this coupling approach can correctly predict the important mechanical phenomena frequently encountered in LSP. This is the second goal of this work which aims to study qualitatively the complex LSP mechanical phenomena. The mechanical loading applied in this study is inspired by those obtained using a specialized laser–matter interaction software. Several interesting materials are routinely used in different laser applications and require additional investigation in their response to laser radiation. Of particular interest is silica glass which is the dominant constitutional material of the optical equipment in laser devices. This material is known to exhibit anomalous behavior in its thermal and mechanical properties [BRÜ 70, BRÜ 71]. Furthermore, certain properties of this glass such as Young's modulus, shear modulus and density show anomalous dependence on the fictive temperature. Because of its complex mechanical behavior, numerical study of this material remains a central issue for several researchers. These reasons have made silica glass an attractive material to be studied numerically. Therefore, it was selected as a part of the LSP application to be studied by the DEM-CM coupling approach. More precisely, the proposed coupling approach will be applied to simulate, from a mechanical point of view, the LSP of silica glass.

I.3. Organization

Following this introduction, the current book is divided into two parts. Part 1 deals with the first objective of this book which is the development of a discrete-continuum coupling approach adapted for highly dynamic multi-scale problems. This part consists of three chapters:

– Chapter 1 reviews some important aspects related to discrete-continuum coupling in dynamics. First, the main coupling challenges to be addressed are detailed. Then, the different concurrent coupling techniques reported in the literature are reviewed. Based on this review, the most appropriate technique is retained to concurrently couple DEM and a CM to be chosen.

– Chapter 2 aims to select the appropriate CM that will be used as a part of the discrete-continuum coupling. First, the most commonly used CMs in computational mechanics are reviewed and classified. Based on this classification, the method that best meets the expectations of the current work is chosen. Finally, the main specificities of the retained method are briefly recalled.

– Chapter 3 focuses on the development of the discrete-continuum coupling method between DEM and the chosen CM. After detailing how the coupling approach is performed, a parametric study of the different coupling parameters is performed. This study aims to draw recommendations simplifying the choice of these parameters in practice, and then to simplify the application of the coupling method on complex problems. Finally, the developed coupling approach is validated using several dynamic reference tests.

Part 2 is dedicated to the application of the developed discrete-continuum coupling approach to qualitatively study the LSP of silica glass, which is the second goal of this book. This part is also divided into three chapters:

– Chapter 4 gives some background knowledge of the different disciplines that interact to accomplish the second objective of this book. First, the laser–matter interaction theory is briefly recalled to identify the important phenomena that must be taken into account to correctly simulate the LSP of silica glass. Then, some important experimental works on the mechanical response of silica glass under different loadings are reviewed. This review serves to provide the main specificities of the silica glass mechanical behavior that is modeled using the proposed coupling approach.

– Chapter 5 focuses on the modeling of silica glass mechanical behavior. A new model intended to faithfully reproduce the different specificities of the silica glass response under highly dynamic loadings is proposed. This model is based on the normal stress in the cohesive beam bonds between discrete elements. Validation of this model is first performed in quasi-statics to simplify the analysis of the potential difficulties, and subsequently in fast dynamics by simulation of high-velocity impact tests of silica glass plates. To describe the silica glass brittle fracture, the virial-stress-based model developed in the first book of this series [JEB 15] is used. The main specificities of this model are briefly recalled at the end of this chapter.

– Chapter 6 investigates the ability of the developed discrete-continuum coupling approach to correctly predict the important mechanical effects characterizing an LSP experiment. To this end, a test of LSP of silica glass is reproduced numerically using this approach as well as the silica glass models detailed in the previous chapter.

Finally, this book ends with several conclusions and outlooks.

PART 1

Discrete-Continuum Coupling Method to Model Highly Dynamic Multi-Scale Problems

1

State of the Art: Concurrent Discrete-continuum Coupling

1.1. Introduction

Some of the most fascinating dynamic problems in all fields of science involve multiple spatial or temporal scales: processes that occur at a certain scale govern the behavior of the system across several (usually coarser) scales. Therefore, to accurately simulate such problems with minimum cost, it is often necessary to resort to multi-scale modeling. In particular, the concurrent discrete-continuum coupling approaches seem to be the best adapted to study these problems. On the one hand, these approaches are of concurrent type. Therefore, as seen previously, these are well adapted to study highly dependent multi-scale phenomena, which are frequently encountered in complex dynamic problems. On the other hand, both discrete and continuum methods (CMs) are involved in these approaches. This allows us to combine their complementary advantages and to avoid their drawbacks. The complex small-scale phenomena can easily be treated by application of the discrete method (DM) in the associated regions. Furthermore, the application of the CM in the remainder of the studied domain significantly reduces the computation time, and avoids prohibitively large computations. Consequently, simulation of real problems of material science and engineering can be performed. Due to their advantages, the concurrent discrete-continuum coupling approaches have become very fashionable in the last decade. This chapter aims to shed light on some important aspects related to these approaches. First, the major challenges that the designer faces in developing such approaches will be outlined. Then, the coupling techniques most commonly used to address these challenges will be reviewed. The one that best meets the scope of the present work will be selected. Although this review is not all-inclusive, it gives a clear vision of the topic. It should be noted that the techniques presented can also be applied for coupling DMs concurrently (concurrent discrete–discrete coupling).

1.2. Coupling challenges

Coupling dissimilar methods in a concurrent manner for dynamic analysis faces two major challenges. The first challenge is due to the different mechanics underlying the DM and CM. The second challenge is related to the different analysis scales of the coupled methods. This section is devoted to explaining these two challenges in more detail.

1.2.1. *Dissimilar variables due to different mechanical bases*

One major difficulty in coupling DM and CM is that these methods are based on different mechanics. The DMs are based on discrete mechanics, in which the interaction between neighboring discrete elements is described by classical Newtonian mechanics. The variables associated with this class of methods are only defined in the element positions. In contrast, the CMs are based on continuum mechanics, in which the studied domain is assumed to be continuous and completely filling the space it occupies. The mechanical behavior of this domain is described by constitutive laws involving continuous field variables, instead of discrete element variables as in the DMs. Therefore, a special treatment is required at the interface between the coupled methods to ensure correct communication between the coupled methods.

1.2.2. *Wave reflections due to different analysis scales*

Another major difficulty encountered in application of the concurrent discrete-continuum coupling approaches in dynamics is that, due to the different analysis scales, spurious wave reflections can occur at the interface between the coupled models. This difficulty is almost always encountered, since there is no benefit in coupling models having similar analysis scales. To simply illustrate the problem of spurious wave reflections, the behavior of waves in the discrete and continuum domains will be explored using simple one-dimensional (1D) models. A chain of particles connected with equivalent springs is used for the discrete model, and its corresponding continuum counterpart is used for the continuum model (Figure 1.1).

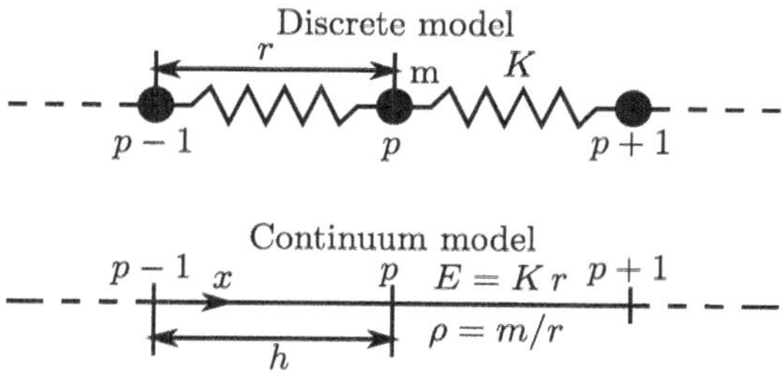

Figure 1.1. *1D models to investigate spurious wave reflections*

In the discrete model, the equation of motion of a particle p, in the absence of body forces, can be derived by application of the fundamental principle of dynamics (FPD):

$$m\,\ddot{d}^p = K\,(d^{p-1} - d^p) + K\,(d^{p+1} - d^p) = K\,(d^{p+1} + d^{p-1} - 2\,d^p) \qquad [1.1]$$

where K is the stiffness of the springs, and m and d^p are, respectively, the mass and displacement of the particle p. Equation [1.1] can be solved by assuming a wavelike solution which is only defined in the particle positions:

$$d^p = D\,e^{i\,(k\,p\,r - \omega\,t)} \qquad [1.2]$$

where D is the amplitude of the oscillations, k is the 1D wave vector (its amplitude gives the wavenumber $\kappa = |k| = \frac{2\pi}{\lambda}$), λ is the wave length, r is the interparticle distance ($x = p\,r$ is the position of the particle p along the chain) and ω is the angular frequency. Substituting [1.2] into [1.1] leads after simplification to:

$$\omega_D^2 = \frac{4\,K}{m} sin^2\left(\frac{k\,r}{2}\right) \qquad [1.3]$$

which is known as the dispersion relation. It characterizes the dependence of the wave frequency on the wavenumber $\kappa = |k|$. It can be shown from [1.3] that $\omega_D(k) = \omega_D(k + \frac{2\,j\,\pi}{r})$ for any $j \in \mathbb{Z}$. Therefore, only the case of $k \in [-\frac{\pi}{r}, \frac{\pi}{r}]$ will be considered hereafter. Moreover, for symmetry reasons, it is sufficient to restrict to $k = \kappa > 0$.

In the continuum model, the 1D equation of motion, in the absence of body forces, can be written as:

$$\rho\,\ddot{u} = \frac{\partial \sigma}{\partial x} \qquad [1.4]$$

where ρ is the density, and σ is the longitudinal Cauchy stress in x direction (Figure 1.1). Under the assumption of small deformations, equation [1.4] can be rewritten as:

$$\rho\,\ddot{u} = E\,\frac{\partial^2 u}{\partial x^2} \qquad [1.5]$$

where E is the Young's modulus. Assuming a harmonic wave propagating along the x direction ($u = U\,e^{i\,(k\,x-\omega\,t)}$), it is easy to find the continuum dispersion relation:

$$\omega_C^2 = \frac{E}{\rho}\,k^2 \qquad [1.6]$$

To ensure equivalence between the discrete and continuum models, E and ρ must verify, respectively, the following conditions (the cross-sectional area S is assumed to be equal to the unit, for simplicity):

$$E = K\,r \quad \text{and} \quad \rho = \frac{m}{r} \qquad [1.7]$$

Using these conditions [1.7], the continuum dispersion relation can be reformulated as follows:

$$\omega_C^2 = \frac{K\,r^2}{m}\,k^2 \qquad [1.8]$$

After obtaining the discrete and continuum dispersion relations ([1.3] and [1.8], respectively), the influence of the spatial and temporal discretization on these parameters will be examined. This allows us to better understand the wave reflection mechanisms encountered when coupling models of different scales.

1.2.2.1. *Influence of the spatial discretization on the dispersion relation*

In this section, the influence of the spatial discretization of the continuum model on the dispersion relation will be examined. To do this, the continuum domain is discretized into uniform segments of length h. Using linear interpolation functions, the discretized equation of motion of a node p can then be expressed as:

$$\rho\, h\, \ddot{u}^p = \frac{E}{h}\,(u^{p-1} - u^p) + \frac{E}{h}\,(u^{p+1} - u^p) = \frac{E}{h}\,(u^{p+1} + u^{p-1} - 2\, u^p) \quad [1.9]$$

where u^p designates the nodal displacement of the node p. By analogy with the discrete model and using [1.7], the dispersion relation of the discretized form of the continuum model can be obtained:

$$\tilde{\omega}_C^2 = \frac{4\, K\, r^2}{m\, h^2} sin^2\left(\frac{k\, h}{2}\right) \quad [1.10]$$

Using [1.3] and [1.10], the cutoff frequencies of the discrete and discretized continuum models can be obtained:

$$f_D^c = \frac{\omega_D^{max}}{2\pi} = \frac{1}{\pi}\sqrt{\frac{K}{m}} \quad [1.11]$$

$$\tilde{f}_C^c = \frac{\tilde{\omega}_C^{max}}{2\pi} = \frac{1}{\pi}\sqrt{\frac{K}{m}}\,\frac{r}{h} \quad [1.12]$$

The cutoff frequency of the discretized continuum model depends on h. In the case of coarse discretization ($h \gg r$), the associated cutoff frequency ($\tilde{f}_C^c$) becomes smaller than that of the discrete one (f_D^c). As a result, if these models are coupled, the high-frequency waves (HFWs) ($f > \tilde{f}_C^c$) coming from the discrete model are not supported by the continuum model, and will be spuriously reflected at the interface (Figure 1.2). This phenomenon has already been addressed using the finite element model with different element sizes [CEL 83].

Taking into account the assumption that $k = \kappa > 0$, the phase and group velocities in both the discrete model and discretized continuum model can be obtained from [1.3] and [1.10] as follows:

$$v_D^{ph} = \frac{\omega_D}{k} = \frac{2}{\kappa}\sqrt{\frac{C}{m}} sin\left(\frac{\kappa\, r}{2}\right) \quad \text{and} \quad v_D^{gr} = \frac{\partial \omega_D}{\partial k} = r\sqrt{\frac{C}{m}} cos\left(\frac{\kappa\, r}{2}\right) \quad [1.13]$$

$$\tilde{v}_C^{ph} = \frac{\omega_C}{k} = \frac{2}{\kappa}\frac{r}{h}\sqrt{\frac{C}{m}}\,sin\left(\frac{\kappa\,h}{2}\right) \quad \text{and}$$

$$\tilde{v}_C^{gr} = \frac{\partial \omega_C}{\partial k} = r\sqrt{\frac{C}{m}}cos\left(\frac{\kappa\,h}{2}\right) \qquad [1.14]$$

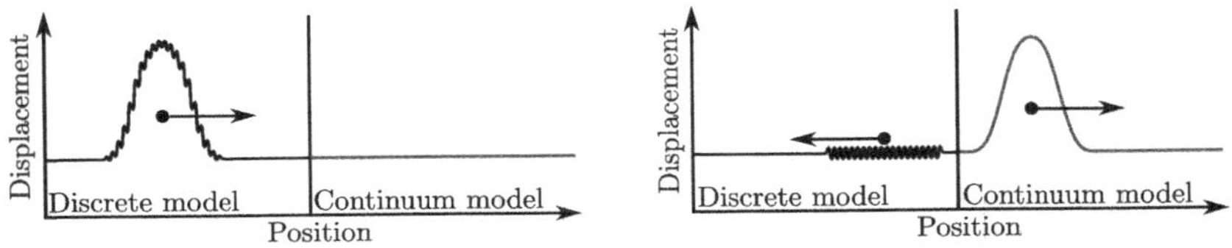

Figure 1.2. *Reflection of high-frequency waves at the interface between the discrete and continuum models*

The use of different discretization characteristic lengths in the discrete and continuum models (r and h) leads to different wave velocities. Moreover, for a fixed κ, the wave velocities in the discretized continuum model decrease as h increases. This explains another mechanism of wave reflections, for which a portion of the main propagating wave is reflected at the interface (not only the HFWs). Figure 1.3 presents examples of a wave traveling between two models having different wave propagation velocities, due to different spatial discretizations.

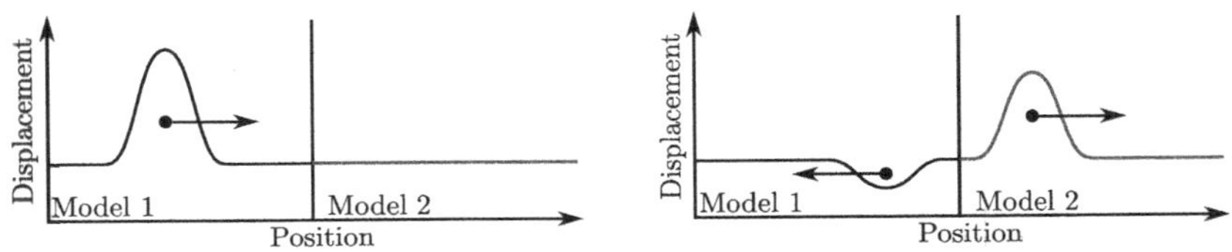

a) Wave traveling from a faster wave velocity model to a slower wave velocity model

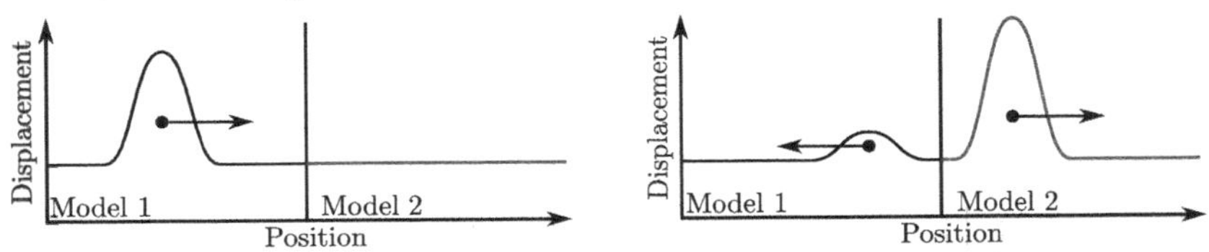

b) Wave traveling from a slower wave velocity model to a faster wave velocity model

Figure 1.3. *Wave traveling between two models having different wave propagation velocities*

1.2.2.2. *Influence of the temporal discretization on the dispersion relation*

As seen before, the use of different discretization characteristic lengths in the discrete and continuum models leads to different dispersion relations, and then to different phase and group velocities. Hereafter, the influence of the temporal discretization on the dispersion relations will be examined. For a given time step Δt, the second derivative of a field variable f with respect to time can be approximated as:

$$\ddot{f} \approx \frac{f_{t+\Delta t} - 2\, f_t + f_{t-\Delta t}}{(\Delta t)^2} \qquad [1.15]$$

Applying [1.15] to the particle and node accelerations ($\ddot{d}^p$ and $\ddot{u}^p$, respectively) and using the equivalence conditions [1.7], equations [1.1] and [1.9] can be approximated as:

$$\frac{m}{\Delta t_D^2}\,(d^p_{n+1} + d^p_{n-1} - 2\, d^p_n) = K\,(d^{p+1}_n + d^{p-1}_n - 2\, d^p_n) \qquad [1.16]$$

$$\frac{m}{\Delta t_C^2}\,(u^p_{n+1} + u^p_{n-1} - 2\, u^p_n) = K\,\frac{r^2}{h^2}\,(u^{p+1}_n + u^{p-1}_n - 2\, u^p_n) \qquad [1.17]$$

where Δt_D and Δt_C are, respectively, the discrete and the continuum time step, and d^p_n and u^p_n are, respectively, the displacement of a particle p and the displacement of a node p at the $n-th$ time step. By analogy with the derivation of the dispersion relations ([1.3] and [1.10]), the following equations can be obtained:

$$sin^2\left(\frac{\omega_D\,\Delta t_D}{2}\right) = \frac{K}{m}\,\Delta t_D^2\, sin^2\left(\frac{k\,r}{2}\right) \qquad [1.18]$$

$$sin^2\left(\frac{\tilde{\omega}_C\,\Delta t_C}{2}\right) = \frac{K}{m}\,\frac{r^2}{h^2}\,\Delta t_C^2\, sin^2\left(\frac{k\,h}{2}\right) \qquad [1.19]$$

Equations [1.18] and [1.19] are obtained using the general form of a harmonic solution which is expressed as $d^p_n = D\, e^{i\,(k\,p\,r - \omega\,n\,\Delta t_D)}$ for a particle p at the $n - th$ time step and $u^p_n = U\, e^{i\,(k\,p\,h - \omega\,n\,\Delta t_C)}$ for a node p at this time step. Using $k = \kappa > 0$, equations [1.18] and [1.19] can be rewritten as:

$$sin\left(\frac{\omega_D\,\Delta t_D}{2}\right) = \sqrt{\frac{K}{m}}\,\Delta t_D\, sin\left(\frac{\kappa\,r}{2}\right) \qquad [1.20]$$

$$sin\left(\frac{\tilde{\omega}_C\,\Delta t_C}{2}\right) = \sqrt{\frac{K}{m}\frac{r}{h}}\,\Delta t_C\,sin\left(\frac{\kappa\,h}{2}\right) \qquad [1.21]$$

With the help of the first-order and third-order Taylor expansions, an approximation of ω_D and $\tilde{\omega}_C$ can be obtained from [1.20] and [1.21] as follows:

$$\omega_D = \sqrt{\frac{K}{m}}\,k\,r\,\left[1 + \frac{k^2\,r^2}{24}\left(\frac{K}{m}\,\Delta t_D^2 - 1\right)\right] \qquad [1.22]$$

$$\tilde{\omega}_C = \sqrt{\frac{K}{m}}\,k\,r\,\left[1 + \frac{k^2\,h^2}{24}\left(\frac{K}{m}\frac{r^2}{h^2}\Delta t_C^2 - 1\right)\right] \qquad [1.23]$$

It can be concluded from the last two equations that, for different discretization characteristic lengths (r and h) and for a given discrete time step Δt_D, it is possible to choose a suitable Δt_C such that the dispersion relations of both models are equivalent:

$$\Delta t_C = \sqrt{\Delta t_D^2 + \frac{m}{K}\,\frac{h^2 - r^2}{r^2}} \qquad [1.24]$$

Therefore, the inaccurate continuum dispersion relation caused by large discretization characteristic length h can theoretically be corrected by choosing a larger continuum time step Δt_C. However, this solution is not always applicable in practice. For example, in the case of a "*dynamic explicit*" simulation, Δt_C must satisfy the Courant–Friedrichs–Lewy (CFL) stability criterion to ensure convergence of numerical approximation [1.25].

$$\Delta t_C \leq \frac{2\,\pi}{\tilde{\omega}_C^{max}} \qquad [1.25]$$

1.3. Coupling techniques

To correctly bridge DM and CM in a concurrent manner for dynamic analysis, the coupling challenges detailed in the previous section should be properly addressed. To this end, several techniques have been proposed in the literature [ABR 98, BRO 99, SMI 99, LU 05, XU 09, JEB 14, CUR 03]. A common feature of these techniques is that the problem domain is often partitioned into several subdomains characterized by different scales and physics. The question that arises here is how to ensure a smooth coupling between these subdomains. The next section tries to answer this question by reviewing the most used techniques.

1.3.1. *Edge-to-edge coupling methods*

A common technique for coupling discrete and continuum models is the edge-to-edge approach. This technique is also widely used in the finite element community for parallel implementation of the finite element method (FEM) method [ZHU 01]. As shown in Figure 1.4, two types of discrete elements are used in this technique. Aside from the real elements, virtual elements are defined to model the angle-bending of the bonds (virtual) between the discrete model and continuum model. The virtual particles (discrete elements) are connected on one side to the discrete model by virtual bonds, and on the other side are attached to the continuum model and move with it. The motion of these particles can be determined by interpolation of the continuum node motion. The most simple variation of this technique is that in which the particles and continuum nodes on the interface are coincident and constrained to move together. This variation is known in the literature as the direct coupling approach. In the general case, the coincidence of particles and nodes on the interface is not necessary. In this case, additional handling is required to ensure correct communication between the coupled models. This last point will be discussed below in more detail.

A domain Ω_G is considered with boundary $\partial\Omega = \partial\Omega^u \cup \partial\Omega^t$ $(\partial\Omega^u \cap \partial\Omega^t = \phi)$, such that essential (displacement) and natural (traction) boundary conditions are, respectively, prescribed on $\partial\Omega^u$ and $\partial\Omega^t$. This domain is divided into two adjacent subdomains Ω_D and Ω_C. These subdomains are, respectively, modeled using the discrete approach and continuum approach (Figure 1.4). An isotropic linear elastic behavior and small deformations are assumed for simplicity. In the present development, an energetic (Hamiltonian) approach will be used to derive the governing equations of the coupling system.

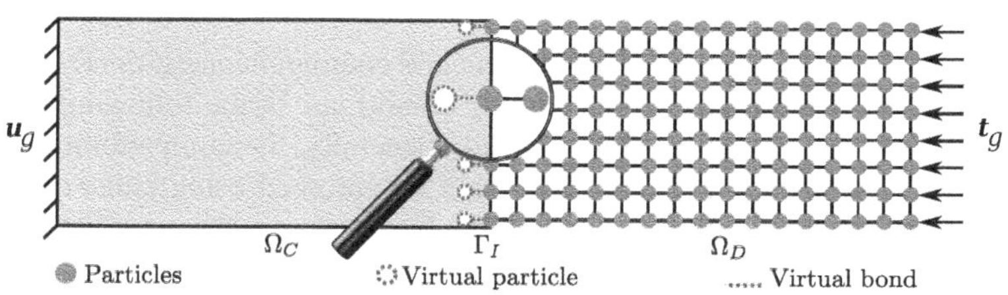

Figure 1.4. *Edge-to-edge coupling*

The total energy in the discrete subdomain Ω_D, which is assumed to be isolated at this stage, is known as the Hamiltonian and is given by:

$$\begin{aligned} H_D &= E_D^{kinetic} + E_D^{int} - W_D^{ext} \\ &= \sum_{i=1}^{n_D} \frac{1}{2\,m_i} \left(\boldsymbol{p}_D^i\right)^2 - \sum_{i=1}^{n_D} \boldsymbol{f}_i^{int}\,\boldsymbol{d}_i + E_D^{virtual} - \sum_{i=1}^{n_D} \boldsymbol{f}_i^{ext}\,\boldsymbol{d}_i \end{aligned} \quad [1.26]$$

where $E_D^{kinetic}$ is the kinetic energy, E_D^{int} is the internal energy, W_D^{ext} is the external work, n_D is the total number of particles, m_i and $\boldsymbol{p}_D^i = m_i\,\dot{\boldsymbol{d}}_i$ are the mass and momentum of the particle i, $\boldsymbol{d}_i$ and $\dot{\boldsymbol{d}}_i$ are, respectively, the displacement and velocity of the particle i, $\boldsymbol{f}_i^{int}$ is the total internal force exerted on the particle i by its neighbors, $\boldsymbol{f}_i^{ext}$ is the total external force acting on the particle i and $E_D^{virtual}$ is the bending energy of the virtual bonds. It should be noted that the stretching energy of the virtual bonds is automatically included in the continuum internal energy, since the virtual particles move with the continuum material. At this stage, $\boldsymbol{f}_i^{ext}$ does not include the coupling force which will be introduced later.

Under the assumption of small deformations, the total energy in the continuum subdomain Ω_C, which is supposed to be isolated at this stage, can be written as:

$$\begin{aligned} H_C &= E_C^{kinetic} + E_C^{int} - W_C^{ext} \\ &= \int_{\Omega_C} \frac{1}{2}\,\rho\,\dot{\boldsymbol{u}}^2\,d\Omega + \tfrac{1}{2}\int_{\Omega_C} \boldsymbol{\sigma} : \boldsymbol{\varepsilon}\,d\Omega - \textstyle\int_{\Omega_C} \rho\,\boldsymbol{b}\,\boldsymbol{u}\,d\Omega - \int_{\partial\Omega_C^t} \boldsymbol{t}_g\,\boldsymbol{u}\,d\Gamma \end{aligned} \quad [1.27]$$

where ρ is the density and $\dot{\boldsymbol{u}}$ is the velocity field, $\boldsymbol{\sigma}$ and $\boldsymbol{\varepsilon}$ are, respectively, the Cauchy stress tensor and strain tensor, $\boldsymbol{b}$ is the body force per unit mass and $\boldsymbol{t}_g$ is the prescribed traction vector on $\partial\Omega_C^t$.

In the above development, the discrete and continuum energies are derived without taking into account the coupling conditions. In the following, it will be shown how these models are coupled. Generally, the compatibility at the interface between the coupled models can be enforced using either velocity constraints or displacement constraints. However, in the case of highly dynamic problems, where dynamic effects become significant, it would be preferable to use velocity constraints to ensure a correct kinetic energy transfer between the coupled models. For the sake of consistency with the energetic (Hamiltonian) approach followed here to develop the edge-to-edge coupling method, displacement constraints will be used in the rest of this derivation. These constraints can be applied in different ways, as shown in [1.28] and [1.29].

$$g_i^{weak} = \|\boldsymbol{u}(\boldsymbol{x}_i) - \boldsymbol{d}_i\|^2 = 0, \quad i \in [1, n_{DI}] \qquad [1.28]$$

$$g_i^{strong} = \boldsymbol{u}(\boldsymbol{x}_i) - \boldsymbol{d}_i = \boldsymbol{0}, \quad i \in [1, n_{DI}] \qquad [1.29]$$

$\boldsymbol{u}$ is the continuum displacement field, $\boldsymbol{d}_i$ is the displacement of a particle i located on the interface at $\boldsymbol{x}_i$ coordinates and n_{DI} is the number of particles on the interface. In [1.28], a single constraint is applied per particle belonging to the interface. This results in a system of equations with a reduced number of degrees of freedom. However, using this type of constraint, displacement compatibility is only satisfied approximately. This can lead to unacceptable large errors. In contrast, three constraints (one for each component) per particle are applied in [1.29]. This allows the coupled models to better communicate. Although it can lead to a large system of equations, [1.29] is generally recommended and is retained here to enforce compatibility between the discrete and continuum approaches. In practice, these constraints can be introduced in the global system by using rigid models, such as the Lagrangian multipliers model (LM), elastic models such as the penalty model (PM) or combined models such as the augmented Lagrange multipliers model (ALM). In the rest of this derivation, the ALM is used, because it is more general and includes the two other models. The associated coupling energy, in a continuous form, is given by:

$$H_I^{ALM} = \int_{\Gamma_I} \boldsymbol{\lambda}\,(\boldsymbol{u} - \boldsymbol{d})\, d\Gamma + \frac{p}{2} \int_{\Gamma_I} (\boldsymbol{u} - \boldsymbol{d})^2\, d\Gamma \qquad [1.30]$$

where Γ_I is the interface between the two models, $\boldsymbol{\lambda}$ is the Lagrange multipliers field and p is the penalty parameter. The Lagrange multipliers field $\boldsymbol{\lambda}$ can be regarded as a generalized coupling force field. Since the displacements $\boldsymbol{d}_i$ in the discrete subdomain Ω_D are only defined in the particle positions, a continuous field $\boldsymbol{d}$ must be inferred from the particle displacements $\boldsymbol{d}_i$ at least on the interface to evaluate H_I^{ALM} [1.30]. This can be achieved using a kind of interpolation:

$$\boldsymbol{d}(\boldsymbol{x}_i) = \sum_{j=1}^{n_{DI}} \phi_D^j(\boldsymbol{x}_i)\, \boldsymbol{d}_j \qquad [1.31]$$

where ϕ_D^j is the interpolation function associated with the particle j. Taking into account [1.30], the total energy of the coupling system can be obtained:

$$H_G^{ALM} = H_D + H_C + H_I^{ALM} \tag{1.32}$$

In [1.32], the Hamiltonian is given in a continuous form. To develop the corresponding discrete form $\tilde{H}_G^{ALM}$, the continuum displacement and the Lagrange multipliers fields ($\boldsymbol{u}$ and $\boldsymbol{\lambda}$, respectively) can be approximated, using, for example, the finite element interpolation in the corresponding domains, as follows:

$$\tilde{\boldsymbol{u}}(\boldsymbol{x}) = \sum_{i=1}^{n_C} \phi_C^i(\boldsymbol{x})\, \boldsymbol{u}_i \tag{1.33}$$

$$\tilde{\boldsymbol{\lambda}}(\boldsymbol{x}) = \sum_{i=1}^{n_I} \phi_I^i(\boldsymbol{x})\, \boldsymbol{\lambda}_i \tag{1.34}$$

where ϕ_C^i and ϕ_I^i are the shape functions constructed on, respectively, $\tilde{\Omega}_C$ and $\tilde{\Gamma}_I$ (the discretized forms of, respectively, Ω_C and Γ_I), n_C and n_I are, respectively, the total number of continuum nodes and the total number of Lagrange multipliers nodes. In the general case, the Lagrange multipliers nodes do not necessarily coincide with the discrete particles or the continuum nodes at the interface. Substituting [1.33] and [1.34] into [1.32] and using [1.31] on Γ_I, the discrete Hamiltonian $\tilde{H}_G^{ALM}$ can be obtained. Based on $\tilde{H}_G^{ALM}$, the discretized Hamiltonian equations of the coupling system can be derived:

$$\dot{\boldsymbol{p}}_D^i = -\frac{\partial \tilde{H}_G^{ALM}}{\partial \boldsymbol{d}_i}, \quad \text{and} \quad \dot{\boldsymbol{d}}_i = \frac{\partial \tilde{H}_G^{ALM}}{\partial \dot{\boldsymbol{p}}_D^i} \quad \text{for } i \in [1,\, n_D] \tag{1.35}$$

$$\dot{\boldsymbol{p}}_C^i = -\frac{\partial \tilde{H}_G^{ALM}}{\partial \boldsymbol{u}_i} \quad \text{and} \quad \dot{\boldsymbol{u}}_i = \frac{\partial \tilde{H}_G^{ALM}}{\partial \dot{\boldsymbol{p}}_C^i} \quad \text{for } i \in [1,\, n_C] \tag{1.36}$$

$$g_i = \frac{\partial \tilde{H}_G^{ALM}}{\partial \boldsymbol{\lambda}_i} = 0 \quad \text{for } i \in [1,\, n_I] \tag{1.37}$$

Replacing $\dot{\boldsymbol{d}}_i$ and $\dot{\boldsymbol{u}}_i$ with their expressions in, respectively, $\dot{\boldsymbol{p}}_D^i$ and $\dot{\boldsymbol{p}}_C^i$, the global system of equations can be expressed, in matrix form, as:

$$\boldsymbol{m}\,\ddot{\boldsymbol{d}} = \boldsymbol{f}^{int} + \boldsymbol{f}^{ext} + \boldsymbol{f}^{\lambda} \tag{1.38}$$

$$\boldsymbol{M}\ddot{\boldsymbol{u}} = \boldsymbol{F}^{int} + \boldsymbol{F}^{ext} - \boldsymbol{F}^{\lambda} \quad [1.39]$$

$$\boldsymbol{C}_I\boldsymbol{u} - \boldsymbol{c}_I\,\boldsymbol{d} = 0 \quad [1.40]$$

where $\boldsymbol{m}$ is the diagonal mass matrix of the particles, $\boldsymbol{f}^{int}$, $\boldsymbol{f}^{ext}$ and $\boldsymbol{f}^{\lambda}$ are the vectors of, respectively, the internal forces, external forces and coupling forces in the discrete model, $\boldsymbol{M}$ is the lumped mass matrix of the nodes, $\boldsymbol{F}^{int}$, $\boldsymbol{F}^{ext}$ and $\boldsymbol{F}^{\lambda}$ are the vectors of, respectively, the internal forces, external forces and coupling forces in the continuum model, $\boldsymbol{C}_I$ and $\boldsymbol{c}_I$ are the continuum and discrete coupling matrices. The matrix systems [1.47], [1.48] and [1.49] are in the form of time-dependent ordinary differential equations (ODEs). These equations can be solved numerically using a time integration scheme.

The edge-to-edge coupling methods are rather dedicated to quasi-static problems or problems with relatively low dynamic effects. Otherwise, fine discretization of the continuum subdomain, at the same level as the discrete subdomain, is required to deal with the spurious wave reflections at the interface. Moreover, in most of the cases, the use of fine continuum discretization is not sufficient to overcome the reflection problems. Mostly, the particles belonging to the interface must be coupled with dampers, which are generally difficult to adjust. These difficulties make the edge-to-edge coupling approaches useless for dynamic studies, since the reduction in the computation effort from a fully discrete element analysis is not significant.

1.3.2. *Bridging domain coupling methods*

In the following, the bridging (overlapping) domain (BD) technique for coupling continuum models with discrete models is described. Contrary to the edge-to-edge coupling technique, the present technique considers a BD (overlapping) between the coupled models. It should be noted that this technique is also called the Arlequin approach in the literature [BEN 98, BEN 01, BEN 05]. It consists of:

– decomposition of the global domain Ω_G into two subdomains Ω_D and Ω_C, modeled, respectively, by the discrete model and continuum model, with a bridging region Ω_B;

– weak coupling (based on weak formulation): the discrepancy between the mechanical states, e.g. displacement, deformation, strain, etc., in the bridging region must be controlled using some kind of fictive forces. To allow each model to express its own wealth, the discrepancy should preferably be controlled in a weak manner using averaging operators. This point consists of:

- definition of a gluing zone Ω_{GL} ($\Omega_{GL} \subseteq \Omega_B$), in which the coupling control will be performed. In the general case, this zone can be different from the bridging region ($\Omega_{GL} \subsetneq \Omega_B$). In the present development, Ω_{GL} is chosen the same that Ω_B. Hereafter, the term "bridging zone" will be used to designate both the bridging zone and gluing zone,

- definition of a mediator space $\mathcal{M}$ which is defined as the space of the field variables restricted to Ω_B that must be controlled in the bridging region. To ensure a correct dialogue between the coupled models, the control quantities must be carefully chosen. Generally, either displacements or velocities are controlled in Ω_B,

- definition of a projection operator Π which will be used to project the discrete and continuum field variables to be controlled onto the mediator space $\mathcal{M}$,

- definition of a junction model which will be used to ensure the compatibility of the controlled field variables in the bridging region. Besides the Lagrange, penalty and augmented Lagrange multipliers models, which were introduced earlier, another junction model has been proposed in the literature [BAU 08, BEN 01, BEN 98, BEN 05, BEN 08]:

$$
\begin{aligned}
< \boldsymbol{\lambda}, \boldsymbol{f}_C - \boldsymbol{f}_D > = \int_{\Omega_B} & \eta_1 \, \boldsymbol{\lambda} \cdot (\Pi \boldsymbol{f}_C - \Pi \boldsymbol{f}_D) \\
& + \eta_2 \, l^2 \varepsilon(\boldsymbol{\lambda}) : \varepsilon(\Pi \boldsymbol{f}_C - \Pi \boldsymbol{f}_D) \, d\Omega
\end{aligned} \qquad [1.41]
$$

where $(\Pi \boldsymbol{f}_C - \Pi \boldsymbol{f}_D)$ is the difference between the projected continuum and discrete control quantities on the mediator space $\mathcal{M}$, $\boldsymbol{\lambda}$ is the Lagrange multiplier field and l is a parameter which is named "*junction parameter*" in this book. This parameter which has the dimension of a length is added to ensure the homogeneity of the integral terms in [1.41]. η_1 and η_2 are non-negative weight parameters. These parameters can be chosen so as to scale the two integral terms in [1.41]. $(\eta_1, \eta_2) = (1, 0)$ refers to $\mathcal{L}^2(\Omega_B)$ inner (scalar) product which is the same that the Lagrange multiplier model, $(\eta_1, \eta_2) = (1, 1)$ refers to $\mathcal{H}^1(\Omega_B)$ inner (scalar) product and $(\eta_1, \eta_2) = (0, 1)$ refers to $\mathcal{H}^1(\Omega_B)$ semi-inner product;

– partition of energy between the discrete and continuum models in the bridging zone. The two models coexist in Ω_B. Therefore, the discrete and continuum energies in this region must be weighted using a kind of unity partition functions (to avoid counting twice the associated energy). Different weight functions can be used for the kinetic energy, internal energy and external work in Ω_B. These functions must verify:

$$
\begin{aligned}
f_C : \Omega_G &\mapsto [0,1] \\
x &\mapsto \begin{cases} 1 & \text{in } \Omega_C \backslash \Omega_B \\ [0,1] & \text{in } \Omega_B \\ 0 & \text{in } \Omega_D \backslash \Omega_B \end{cases}
\end{aligned}
\qquad [1.42]
$$

in the continuum subdomain, and $f_D = \bar{f}_C = 1 - f_C$ in the discrete subdomain. Figure 1.5 presents examples of weight functions.

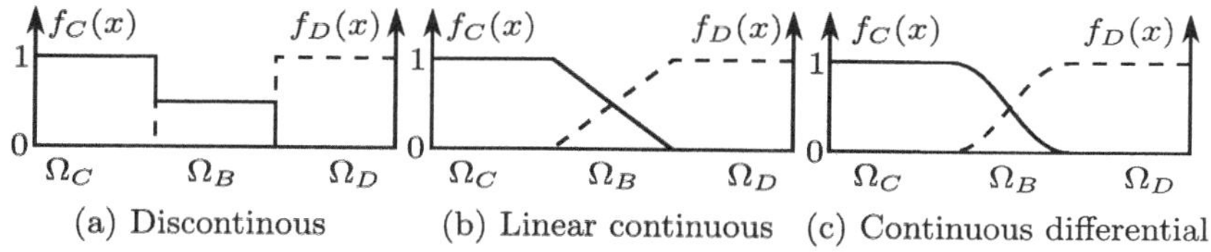

Figure 1.5. *Examples of weight functions*

These different ingredients will be detailed hereafter. As for the edge-to-edge technique, a domain Ω_G is considered with boundary $\partial\Omega = \partial\Omega^u \cup \partial\Omega^t$ ($\partial\Omega^u \cap \partial\Omega^t = \phi$), such that essential (displacement) and natural (traction) boundary conditions are, respectively, prescribed on $\partial\Omega^u$ and $\partial\Omega^t$. This domain is divided into two subdomains Ω_D, treated by the discrete model, and Ω_C, treated by the continuum model, with a bridging region Ω_B (Figure 1.6). An isotropic linear elastic behavior and small deformations are assumed for simplicity.

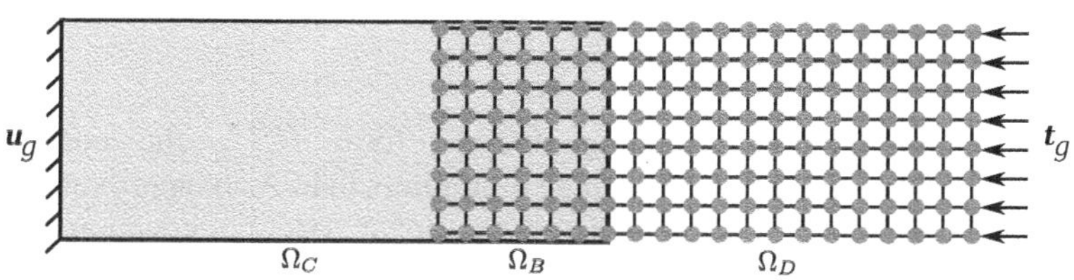

Figure 1.6. *Bridging domain coupling*

To weight the energies in the continuum subdomain, three weight functions are assumed in this development: α for the internal energy, β for the kinetic energy and γ for the external work. In a complementary manner, the energies in the discrete subdomain are weighted using $\bar{\alpha}$, $\bar{\beta}$ and $\bar{\gamma}$, for, respectively, the internal energy, kinetic energy and external work. Using these functions, the weighted discrete and continuum Hamiltonians can, respectively, be written as:

$$\begin{aligned} H_D^w &= E_D^{kinetic,\, w} \qquad\qquad + E_D^{int,\, w} \qquad\qquad - W_D^{ext,\, w} \\ &= \sum_{i=1}^{n_D} \frac{1}{2\, m_i}\, \bar{\beta}_i\, \left(\boldsymbol{p}_D^i\right)^2 - \sum_{i=1}^{n_D} \bar{\alpha}_i\, \boldsymbol{f}_i^{int}\, \boldsymbol{d}_i - \sum_{i=1}^{n_D} \bar{\gamma}_i\, \boldsymbol{f}_i^{ext}\, \boldsymbol{d}_i \end{aligned} \qquad [1.43]$$

$$\begin{aligned} H_C^w &= E_C^{kinetic,\, w} + E_C^{int,\, w} - W_C^{ext,\, w} \\ &= \int_{\Omega_C} \frac{1}{2}\, \beta\, \rho\, \dot{\boldsymbol{u}}^2\, d\Omega + \frac{1}{2} \int_{\Omega_C} \alpha\, \boldsymbol{\sigma} : \boldsymbol{\varepsilon}\, d\Omega \\ &\quad - \int_{\Omega_C} \gamma\, \rho\, \boldsymbol{b}\, \boldsymbol{u}\, d\Omega - \int_{\partial\Omega_C^t} \gamma\, \boldsymbol{t}_g\, \boldsymbol{u}\, d\Gamma \end{aligned} \qquad [1.44]$$

To enforce compatibility between the coupled models in the bridging zone Ω_B, the $\mathcal{H}^1(\Omega_B)$ junction model ([1.41], with $(\eta_1, \eta_2) = (1, 1)$) is used. The associated coupling energy can be written as:

$$H_B^{H1} = \int_{\Omega_B} \boldsymbol{\lambda} \cdot (\Pi \boldsymbol{u} - \Pi \boldsymbol{d}) + l^2 \varepsilon(\boldsymbol{\lambda}) : \varepsilon(\Pi \boldsymbol{u} - \Pi \boldsymbol{d})\, d\Omega \qquad [1.45]$$

As explained before, to evaluate H_B^{H1}, a continuous displacement field $\boldsymbol{d}$ must be approximated from the particle displacements $\boldsymbol{d}_i$ in the bridging zone, using a kind of interpolation [1.31]. The global Hamiltonian of the coupling system can be obtained by summing [1.43], [1.44] and [1.45]:

$$H_G^{H1} = H_D^w + H_C^w + H_B^{H1} \qquad [1.46]$$

To obtain the corresponding discrete form $\tilde{H}_G^{H1}$, the continuum displacement field $\boldsymbol{u}$ and the Lagrange multipliers field $\boldsymbol{\lambda}$ are approximated as given by [1.33] and [1.34]. Using $\tilde{H}_G^{H1}$, the global system of equations can be derived:

$$\boldsymbol{m}_\beta\, \ddot{\boldsymbol{d}} = \boldsymbol{f}_\alpha^{int} + \boldsymbol{f}_\gamma^{ext} + \boldsymbol{f}^\lambda \qquad [1.47]$$

$$\boldsymbol{M}_\beta\, \ddot{\boldsymbol{u}} = \boldsymbol{F}_\alpha^{int} + \boldsymbol{F}_\gamma^{ext} - \boldsymbol{F}^\lambda \qquad [1.48]$$

$$\boldsymbol{C}_B\, \boldsymbol{u} - \boldsymbol{c}_B\, \boldsymbol{d} = 0 \qquad [1.49]$$

where $\boldsymbol{m}_\beta$ is the weighted diagonal mass matrix of the particles, $\boldsymbol{f}_\alpha^{int}$, $\boldsymbol{f}_\gamma^{ext}$ and $\boldsymbol{f}^\lambda$ are, respectively, the weighted vector of internal forces, weighted vector of external forces and vector of coupling forces in the discrete model, $\boldsymbol{M}_\beta$ is the weighted continuum lumped mass matrix, $\boldsymbol{F}_\alpha^{int}$, $\boldsymbol{F}_\gamma^{ext}$ and $\boldsymbol{F}^\lambda$ are, respectively, the weighted vector of internal forces, weighted vector of the external forces and vector of coupling forces in the continuum model. $\boldsymbol{C}_B$ and $\boldsymbol{c}_B$ are the continuum and discrete coupling matrices. For $\boldsymbol{M}_\beta$ and $\boldsymbol{m}_\beta$ to be invertible, the weight functions β and $\bar{\beta}$ must be strictly positive in Ω_B and at the border $\partial\Omega_B$. Therefore, a small ε must be used instead of zero in the nodes assigned to $\partial\Omega_B$. The definition of the weight function β given by [1.42] is slightly modified as follows:

$$\begin{aligned} \beta : \Omega_G &\rightarrow [0,1] \\ x &\rightarrow \begin{cases} 1 & \text{in } \Omega_C \backslash \Omega_B \\ [\varepsilon, 1-\varepsilon] & \text{in } \Omega_{C|\Omega_B} \\ 0 & \text{in } \Omega_D \backslash \Omega_B \end{cases} \end{aligned} \qquad [1.50]$$

The global system of equations consisting of [1.47], [1.48] and [1.49] can finally be solved using a temporal integration scheme.

In the BD coupling methods, the fine-scale solution (of the discrete model) is continuously projected onto the coarse-scale solution (of the continuum model) in the bridging zone. This can cancel the HFWs, and then avoids spurious wave reflections at the discrete/continuum interface, without any additional filtering or damping [BEN 05, JEB 13c]. However, this is conditional upon choosing correctly the coupling parameters, which is not a straightforward issue due to the large number of these parameters. Another interesting aspect of the BD coupling methods is that they can also be applied to obtain solution for nonlinear problems, in which devising an energy functional for the entire system is not possible, due to the presence of irreversible process, for example [FIS 07].

1.3.3. *Bridging-scale coupling methods*

The bridging-scale technique was recently proposed by Wagner and Liu [WAG 03] to couple molecular dynamics (MD) and FEM. The main concepts of this approach will be briefly reviewed hereafter. The readers can refer to [PAR 05a, PAR 05b, WAG 03] for more details. Two subdomains Ω_D and Ω_C modeled, respectively, by discrete and continuum models are considered. These will first be assumed to be completely superimposed in the global domain Ω_G, to better understand the features of the bridging-scale approach.

The key idea of such an approach is to decompose the total displacement field $\boldsymbol{u}$ in the particle positions into coarse and fine scales:

$$\boldsymbol{u}_{Tot}(\boldsymbol{x}_i) = \bar{\boldsymbol{u}}(\boldsymbol{x}_i) + \boldsymbol{u}'(\boldsymbol{x}_i) \quad [1.51]$$

where $\boldsymbol{x}_i$ represents the position of a particle i, $\bar{\boldsymbol{u}}$ and $\boldsymbol{u}'$ are, respectively, the coarse-scale and fine-scale displacements (Figure 1.7).

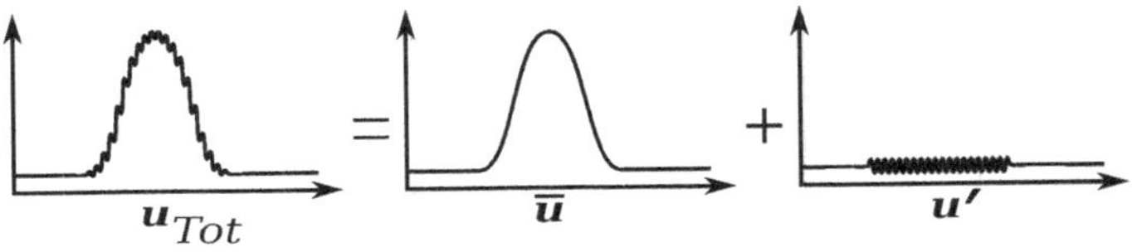

Figure 1.7. *Scale decomposition of total displacement in the particle positions*

After discretization of the continuum subdomain Ω_C, the coarse-scale displacement in the particle positions $\boldsymbol{x}_j$ is evaluated from the continuum nodal displacements as:

$$\bar{\boldsymbol{u}}(\boldsymbol{x}_j) = \sum_{i=1}^{n_C} \phi_C^i(\boldsymbol{x}_j)\, \boldsymbol{u}_i \quad [1.52]$$

where $\boldsymbol{u}_i$ and ϕ_C^i are the continuum nodal displacement and shape function associated with the node i, respectively. For the sake of clarity, matrix representation will be used for subsequent developments. In matrix form, equation [1.52] can be rewritten as:

$$\bar{\boldsymbol{u}} = \boldsymbol{N}\, \boldsymbol{u} \quad [1.53]$$

where $\bar{\boldsymbol{u}} = {}^t\{\bar{\boldsymbol{u}}(\boldsymbol{x}_1)\; \bar{\boldsymbol{u}}(\boldsymbol{x}_2) \;...\; \bar{\boldsymbol{u}}(\boldsymbol{x}_{n_D})\}$ is the vector of coarse displacements in the particles positions, n_D is the total number of particles, $\boldsymbol{u} = {}^t\{\boldsymbol{u}_1\; \boldsymbol{u}_2 \;...\; \boldsymbol{u}_{n_C}\}$ is the vector of continuum nodal displacements, n_C is the total number of nodes and $\boldsymbol{N}$ is the interpolation matrix.

The fine-scale displacement is defined as the part of the total displacement that cannot be represented by the coarse scale. This quantity is computed from the vector $\boldsymbol{d}$ of the particle displacements, which include the coarse-scale parts, by subtracting (from $\boldsymbol{d}$) the projection of the discrete solution onto the continuum solution.

$$\begin{aligned} \boldsymbol{u}' &= \boldsymbol{d} - \boldsymbol{P}\,\boldsymbol{d} \\ &= \boldsymbol{Q}\,\boldsymbol{d} \end{aligned} \qquad [1.54]$$

where $\boldsymbol{P}$ is the projection matrix and $\boldsymbol{Q} = \boldsymbol{I} - \boldsymbol{P}$ is the complementary projector [WAG 03]. The term $\boldsymbol{P}\,\boldsymbol{d}$, which is called "bridging scale (BS)", is the part of the particle displacement that must be subtracted from the total displacement to completely separate the scales (i.e. the coarse and fine scales are orthogonal or linearly independent of each other). The total displacement $\boldsymbol{u}$ can finally be written as the sum of the coarse and fine scales as:

$$\boldsymbol{u}_{Tot} = \boldsymbol{N}\,\boldsymbol{u} + \boldsymbol{Q}\,\boldsymbol{d} \qquad [1.55]$$

To derive the coupled discrete and continuum equations, it is most convenient to adopt the Lagrangian approach, in which the multi-scale Lagrangian is defined by:

$$\begin{aligned} L(\boldsymbol{u}_{Tot}, \dot{\boldsymbol{u}}_{Tot}) &= E^{kinetic}(\dot{\boldsymbol{u}}_{Tot}) - E^{int}(\boldsymbol{u}_{Tot}) + W^{ext}(\boldsymbol{u}_{Tot}) \\ &= \tfrac{1}{2}\,{}^{t}\dot{\boldsymbol{u}}\,\boldsymbol{M}\,\dot{\boldsymbol{u}} + \tfrac{1}{2}\,{}^{t}\dot{\boldsymbol{d}}\,\boldsymbol{m}_f\,\dot{\boldsymbol{d}} - E^{int}(\boldsymbol{u}_{Tot}) +^{t} \boldsymbol{f}^{ext}\,\boldsymbol{u}_{Tot} \end{aligned} \qquad [1.56]$$

where $\boldsymbol{M}$ and $\boldsymbol{m}_f$ are, respectively, the coarse-scale and fine-scale diagonal mass matrices (defined in terms of the diagonal mass matrix $\boldsymbol{m}$ of the particles as: $\boldsymbol{M} = \boldsymbol{N}^t\,\boldsymbol{m}\,\boldsymbol{N}$ and $\boldsymbol{m}_f = \boldsymbol{Q}^t\,\boldsymbol{m}\,\boldsymbol{Q} = \boldsymbol{Q}^t\,\boldsymbol{m} = \boldsymbol{m}\,\boldsymbol{Q}$), and $\boldsymbol{f}^{ext}$ is the vector of external forces acting on the particles. An important feature of the Lagrangian [1.56] is the absence of quadratic terms in the kinetic energy, which are canceled due to the presence of the "BS". The coupled multi-scale equations of motion can be obtained from L as follows:

$$\frac{d}{dt}\left(\frac{\partial L}{\partial \dot{\boldsymbol{u}}}\right) - \frac{\partial L}{\partial \boldsymbol{u}} = 0 \quad \text{and} \quad \frac{d}{dt}\left(\frac{\partial L}{\partial \dot{\boldsymbol{d}}}\right) - \frac{\partial L}{\partial \boldsymbol{d}} = 0 \qquad [1.57]$$

Using [1.56], these equations can be rewritten after simplification as:

$$\boldsymbol{m}\ddot{\boldsymbol{d}} = \boldsymbol{f}^{int} + \boldsymbol{f}^{ext} \qquad [1.58]$$

$$\boldsymbol{M}\ddot{\boldsymbol{u}} = \boldsymbol{N}^t\left(\boldsymbol{f}^{int} + \boldsymbol{f}^{ext}\right) \qquad [1.59]$$

The first (fine scale) equation is the equation of motion in the discrete model, $\boldsymbol{m}$ is the diagonal mass matrix of the particles. The second (coarse scale) equation is simply the equation of motion in the continuum model, where the mass matrix $\boldsymbol{M}$ is consistent. The coupling between these equations is ensured through the vector of the continuum internal forces $\boldsymbol{F}^{int} = \boldsymbol{N}^t \boldsymbol{f}^{int}$ which is a function of the vector of the discrete internal forces $\boldsymbol{f}^{int}$.

In the above development, Ω_D and Ω_C are assumed to be completely superimposed. In this case, the continuum equation of motion [1.59] is redundant, since it is only an approximation of the discrete one [1.58]. The total and discrete displacements ($\boldsymbol{u}_{Tot}$ and $\boldsymbol{d}$, respectively) satisfy the same equation of motion [1.58] and have the same initial conditions. Therefore, these quantities are identical during the entire simulation, and the continuum displacement $\boldsymbol{u}$ can simply be determined by [1.55] which implies: $\boldsymbol{N}\,\boldsymbol{u} = \boldsymbol{P}\,\boldsymbol{d}$.

Assuming that the fine-scale analysis is only required in a small region Ω_B, the particles outside this zone will now be removed. In other words, the unnecessary fine-scale degrees of freedom will be eliminated. By doing so, the discrete model is reduced to Ω_B, where the two models coexist, and the remaining estate is only modeled by the continuum model (Figure 1.8). The process of eliminating the unnecessary degrees of freedom results in a modified equation of motion in the discrete model, including an external force called impedance force. This force is a function of a damping matrix or equivalently its time derivative, known as the time history kernel [ADE 76, WAG 03]. The damping kernel was first derived analytically by Adelman and Doll [ADE 76] for a harmonic 1D lattice. However, their analytical approach is very costly and is intractable above 1D [PAR 05b]. To overcome this limitation, other works [CAI 00, WEI 02] have proposed numerical approaches to evaluate this quantity, whose the effect is to dissipate the HFWs and to avoid wave reflections at the interface between the discrete and continuum models.

In recent years, the bridging-scale methods have received much research interest, and several improvements, concerning the damping kernel and projection mechanism, have been proposed. In the first implementation of Wagner and Liu [WAG 03], the projection of the fine-scale solution onto the coarse-scale solution is apparently only performed at the interface using the approach of Adelman and Doll [ADE 76] (without introducing Lagrange multipliers). This approach has been improved by computing the interface nodes using Green's functions for lattices [PAR 05a, PAR 05b]. More recently, other enhancements have been proposed to join the coarse and fine scales in a BD, using the perfectly matched layer (PML) technique

[LI 06, TO 05]. As shown [XU 09], the performance of the BS methods based on PML is comparable to that of BD methods. However, the BD methods would be less costly, since they allow for reducing spurious wave reflections by only a correct setting of their parameters (no additional treatments are required).

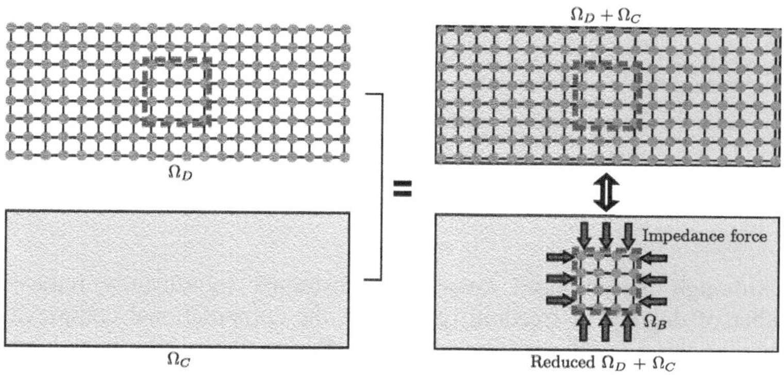

Figure 1.8. *Removal of unnecessary degrees of freedom in bridging-scale modeling*

1.3.4. *Other coupling techniques*

1.3.4.1. *Quasicontinuum method*

The quasicontinuum (QC) method was originally developed in the context of lattice statics at zero temperature (molecular mechanics) using empirical interatomic potentials [TAD 96b, TAD 96a]. Later on, it was extended to dynamic problems using the coarse grained energy and the Hamilton principle [LI 14, ROD 03, SHE 99]. The chief objective of the theory is to systematically coarsen a particle (atomistic) description by introducing kinematic constraints. These constraints are selected and designed so that the fully atomistic model is preserved in the regions of fine-scale effects and large number of particles (atoms) are collectively treated in the coarse-scale regions. The fully discrete model is then represented by a set of representative particles (mainly located in the fine-scale region), where the equation of motion has to be solved (Figure 1.9). The displacement of the non-representative particles is determined by kinematic constraints based on finite element meshing, of which the nodes are coincident with the representative particles (Figure 1.9).

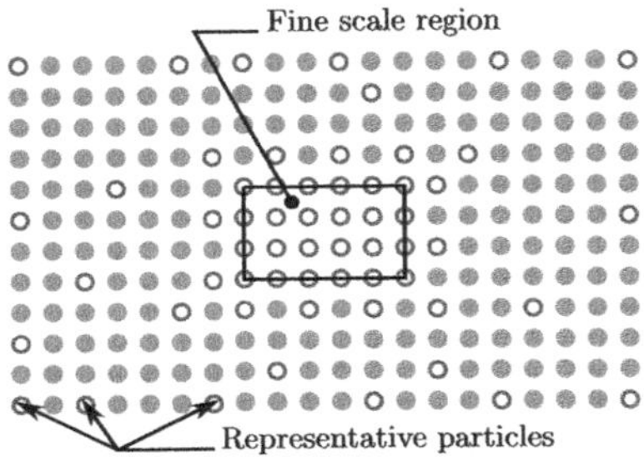

a) Selection of representative particles

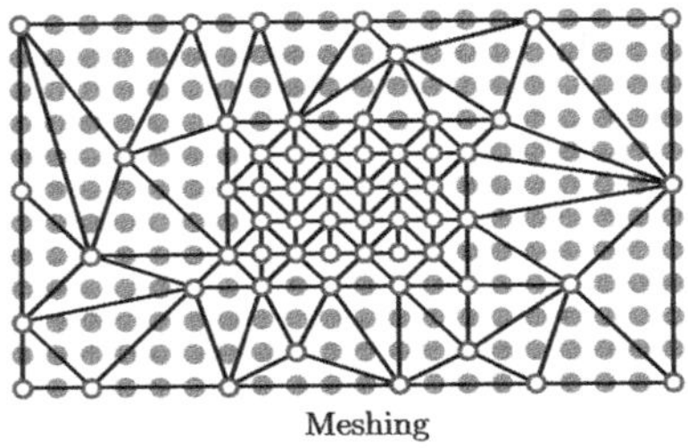

b) Meshing for application of kinematic constraints

Figure 1.9. *Quasicontinuum model*

Although the imposed kinematic constraints significantly reduce the number of degrees of freedom (particles), the computational complexity of evaluating the generalized forces corresponding to the coarse-grained degrees of freedom (representative particles) still scales with the total number of particles in the system making computation on large systems intractable. Several approximations have been suggested in the literature to alleviate such difficulty [IYE 11, LI 14, ORT 01, ROD 03, SHE 99]. These include the mixed discrete (atomistic) and continuum formulations, or introduction of cluster summation rules on lattice sums. However, these approximations induce spurious forces, which can affect the solution accuracy. As shown in the literature [LI 14], the effects of these induced forces are more significant in the case of dynamic studies. These effects may even be more severe than the spurious reflections at the interface. Many strategies have been suggested to correct the errors incurred in these approximations [SHI 04, WEI 06], but they introduce undesirable seams in the process. Furthermore, recent numerical analysis suggests that the approximations introduced may not be consistent and stable, and can result in uncontrolled errors for rapid coarse-graining [DOB 08, DOB 10]. More recently, other works have proposed seamless QC formulations. However, they generally suffer from a lack of systematic convergence [IYE 11].

1.3.4.2. *Coupling of discrete and continuum meshless methods*

In this section, techniques used to couple discrete and continuum meshless methods are briefly reviewed. The readers are referred to [LIU 02] for more details. These techniques were first developed to couple smoothed particle hydrodynamics (SPH) and MD. In meshless methods, the studied domain is represented by a set of scattered particles (continuum particles) or nodes without any connectivity between them. Each continuum particle has its corresponding smoothing length R, representing the influence domain, and length scale h, characterizing locally the domain discretization. The idea is that the continuum particles close to the discrete-continuum interface (transitional particles) also act as virtual discrete particles for the discrete model (Figure 1.10). To avoid interface problems, the length scale of the transitional particles must be graded down to the order of the discretization characteristic length of the discrete model. The communication between the coupled models can further be enhanced by using large transitional zone, in which the particles act both as continuum and discrete particles (Figure 1.10).

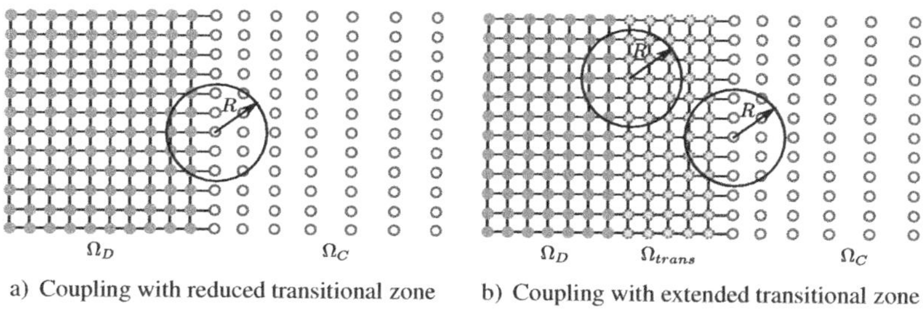

Figure 1.10. *Techniques for coupling discrete and continuum meshless methods*

1.4. Conclusion

Mainly, three coupling approaches can be used to couple discrete element method (DEM) with CMs: hierarchical, concurrent and hybrid hierarchical-concurrent approach. Among them, the concurrent approach is the most appropriate to study multi-scale problems in fast dynamics which is the scope of the present book. This chapter brought light to some important aspects related to this approach. First, the major difficulties that arise from application of this approach to couple DM and CM for dynamic studies were briefly outlined. These difficulties are mainly due to the different mechanical bases and analysis scales between the coupled methods. Then, the techniques most commonly used to overcome such difficulties were reviewed. Among these techniques, the BD technique seems best suited to develop a concurrent DEM-CM coupling approach for highly dynamic studies. Using this technique, spurious wave reflections can naturally be avoided without any additional filtering or damping. This can considerably reduce the computation time. The fine-scale solution is projected onto the coarse-scale solution in the bridging region at each time step. This projection mechanism naturally filters the high-frequency portion of the fine-scale solution that is not supported by the admissible solution space of the coarse-scale model. This technique is then retained to develop the concurrent discrete-continuum coupling. The question that arises here is how to choose the CM that will be coupled with DEM. Chapter 2 tries to answer this question.

2

Choice of the Continuum Method to be Coupled with the Discrete Element Method

2.1. Introduction

The continuum methods (CMs), such as the finite element method (FEM) [ZIE 05a, ZIE 05b, ZIE 05c] and smoothed particle hydrodynamics (SPH) [LIU 03, LUC 77], are used to study macroscopic physical phenomena having length scales much greater than the interatomic distance. This class of method is well adapted for continuous problems, of which the continuity assumption is valid and remains valid during calculation. Otherwise, additional treatment must be considered to deal with discontinuities. Solving mechanical problems using CMs is typically a two-step process. In the first step, the problem domain which is assumed to be continuous is discretized into a finite number of discrete components made up of reference points or nodes. Then, the continuous governing equations obtained from continuum mechanics are converted into a system of algebraic equations easy to solve. Solving this system gives the solutions at the reference points. For other points in the occupied space, the field variables can be approximated by interpolation or averaging (in the case of SPH) of the reference point solutions.

Although they are less suitable to study complex phenomena, CMs are generally much less expensive than the discrete methods, which explains the great scientific interest in them. This has resulted in the development of a wide variety of CMs, which sometimes makes it difficult to choose one of them for a given application. Particularly, the choice of the CM that will be coupled with the discrete element method (DEM) is not straightforward. To simplify this issue, it would be helpful to divide the CMs into several classes; each class includes the methods that share the main features in their principle. By doing so, a first classification of the classes (not methods) can be preformed to select the class that best complies with the objectives of this work. Then, the requested CM can be selected from only the retained class. This is the purpose of the present chapter which, after selecting the CM, will review its main specificities and features.

2.2. Classification of the continuum methods

The existing CMs in the literature can mainly be divided into two classes: grid-based methods and meshless methods (Figure 2.1). The main features of each class are reviewed hereafter.

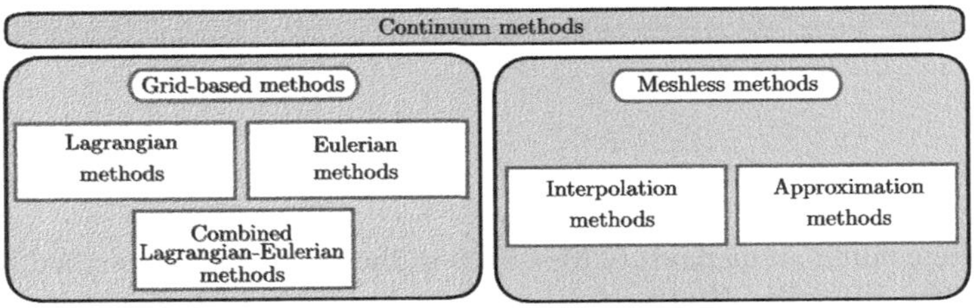

Figure 2.1. *Classification of continuum methods*

2.2.1. *Grid-based methods*

In the grid-based methods, the spatial domain is often represented by discretized elements. These elements are called "meshes" in the case of FEM [ZIE 05c], grids in the case of the finite difference method (FDM) [SMI 85] and volumes or cells in the case of the finite volume method (FVM) [VER 07]. The terminologies of grids, volumes, cells and elements carry different physical meanings linked to physical problems. However, all of them can be called meshes according to the following definition. A mesh, in the broadest sense of the word, is defined as an open space or interstice between the strands of a network formed by connecting nodes in a predefined manner. The key here is that, in the grid-based methods, the mesh must be predefined to provide a certain relationship between the nodes. This is the formulation basis of this class of methods. Based on a properly predefined mesh, complex ordinary differential or partial differential governing equations can be approximated by a set of algebraic equations for each discretized element (mesh). The system of algebraic equations for the whole problem domain can be formed by assembling the elementary algebraic equations for all the meshes.

There are two main frameworks for describing the governing equations of mechanical problems: the Lagrangian description and Eulerian description. The former, also called the material description, is a way to look at material motion where the observer follows an individual material point as it moves through the space and time. The latter, also called the spatial description, is another way to look at material motion where the observer focuses on a specific space position through which the material passes over time. The use of one or other description leads to quite different governing equations. The difference between them arises from the definition of the *total time derivative* which is defined, in the case of Eulerian description, as the sum of the *local time derivative* and the *convective derivative* [2.1]:

$$\underbrace{\frac{d}{dt}}_{total\ time\ derivative} = \underbrace{\frac{\partial}{\partial t}}_{local\ time\ derivative} + \underbrace{\boldsymbol{V}.\nabla}_{convective\ derivative} \qquad [2.1]$$

where $\boldsymbol{V}$ represents a velocity vector and ∇ denotes the vector differential operator (gradient). According to the used description, the grid-based methods can be divided into three subclasses: Lagrangian methods, Eulerian methods and combined Lagrangian–Eulerian methods.

2.2.1.1. *Lagrangian methods*

Lagrangian methods which are typically represented by FEM [ZIE 05c] are based on the Lagrangian description. In this class of methods, the grid or mesh is attached to the material in the entire computation process, and moves with it (Figure 2.2).

Each node in the mesh follows the material path at the grid point. As a result, the mesh elements can be deformed during the simulation process. When the domain deforms, the mesh elements deform accordingly. Mass, momentum and energy are transported with the motion of the mesh. Because the mass within each cell (or element) remains constant, no mass flux is allowed through the mesh cells boundaries. The Lagrangian methods present several advantages:

– it is very easy to treat the boundary conditions at free surfaces, moving boundaries and material interfaces. Indeed, they are automatically imposed, tracked and determined simply by placing some grid nodes along the boundaries and material interfaces;

– complex geometries can be conveniently studied using an irregular mesh;

– meshing is only required within the problem domain (no additional grids are required beyond it, as will be seen for Eulerian methods). Therefore, Lagrangian methods are computationally efficient;

– since the mesh is attached to the material, the entire time history of all field variables at a specific material point can easily be obtained during the simulation;

– in the Lagrangian description, the *total time derivative* is the same as the *local time derivative*. Therefore, no convective terms exist in the associated governing equations. This makes Lagrangian simulations simpler and faster as no computational effort is required to treat the convective terms.

Figure 2.2. *Lagrangian mesh for a rabbit simulation*

Due to these advantages, Lagrangian methods have become very popular to simulate problems of computational solid mechanics (CSMs), where deformations are relatively small. However, application of these methods to solve problems with large deformations presents a huge challenge. Indeed, in such problems, the mesh can become extremely distorted, and stability as well as convergence difficulties can arise, leading to computation breakdown. The accuracy of the formulation and consequently the solution can be severely affected. Also, the time step which is controlled by the size of the smallest element grid can become too small to be efficient for computation.

A solution to overcome these problems is to re-mesh (rezone) the problem domain or simply the regions where the initial mesh is hugely distorted. This technique involves overlying of a new (undistorted) mesh on the old one. The computation is then resumed on the new mesh. The field variables can be approximated at the new grid nodes via mass, momentum and energy transport using an Eulerian description. Despite the popularity of the rezoning techniques to simulate large deformation problems, they suffer from several difficulties. Indeed, rezoning procedure can be tedious and time-consuming. Besides, the transport of the field variables from old to new mesh is generally accompanied by material diffusion which can lead to a loss of the material history [BEN 92].

2.2.1.2. *Eulerian methods*

In Eulerian methods, the mesh is fixed on the space occupied by the simulated object which can move across the fixed mesh cells (grids) as shown in Figure 2.3. All the mesh cells as well as the associated nodes remain fixed in the space and do not move with the material moving across the mesh. The flux of mass, momentum and energy across mesh cell boundaries is simulated to compute the distribution of different field variables in the problem domain. As a result, the shape and volumes of the grid cells remain constant during the entire computation process. Therefore, simulation of problems with large deformations is possible with Eulerian methods. Since there is no cells deformation, these methods do not cause the same kind of numerical problems as in the Lagrangian methods. For these reasons, Eulerian methods are dominant in computational fluid dynamics (CFD) problems and problems with large deformations such as explosion, high-velocity impacts (HVIs), etc.

Despite the great success of Eulerian methods, there are many disadvantages which limit their application, from which we can cite:

– since the Eulerian methods simulate the flux of mass, momentum and energy across mesh cells boundaries, it is very difficult to get out the time history of field variables at a fixed material point;

– Eulerian methods require meshing not only within the problem domain, but also beyond it (Figure 2.3). The mesh must be large enough to cover the area through which the simulated material can move. Therefore, coarse grid is sometimes recommended for computational efficiency, which can affect the solution accuracy;

– the free surfaces, moving and deformable boundaries and moving material interfaces are difficult to determine accurately;

– it is difficult to treat irregular and complicated geometries of materials. Generally, tedious mesh generation procedure to convert the irregular geometry of the problem domain into a regular computational domain is necessary. This can become sufficiently expensive to be efficient. Efforts in this direction are still ongoing.

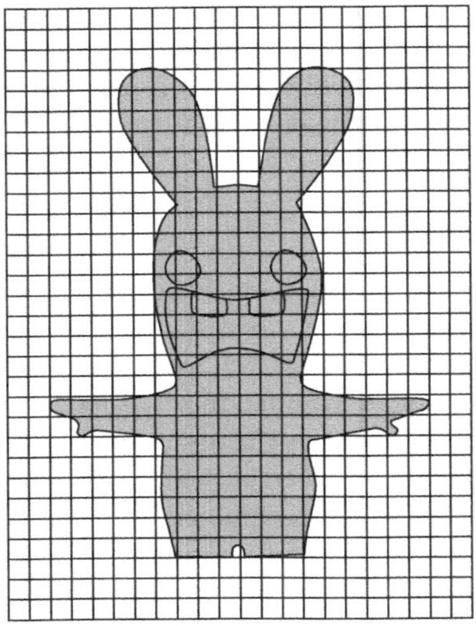

Figure 2.3. *Eulerian mesh for a rabbit simulation*

2.2.1.3. *Combined Lagrangian–Eulerian methods*

The features of the Lagrangian and Eulerian methods are largely complementary. This has led to the development of several approaches involving the Lagrangian and Eulerian descriptions at the same time. The coupled Eulerian Lagrangian (CEL) [FEL 92, MAI 95, XIE 09] and the arbitrary Lagrangian Eulerian (ALE) [BEN 92, STE 04] can particularly be cited. The CEL approach consists of simultaneously employing both the descriptions in separate (or with overlap) regions of the problem domain. The Lagrangian region continuously interacts with the Eulerian region through a coupling module allowing for exchange of computational information. The ALE approach is conceptually close to the rezoning techniques for Lagrangian methods. In this case, the mesh moves independently of the material so as to minimize the mesh distortion. In the ALE approach, Lagrangian motion is firstly computed at each time step, followed by possible rezoning step in which the mesh is either rezoned to the original shape or rezoned to some more advantageous shape (between the previous shape and the currentshape).

The combined Lagrangian–Eulerian methods are very promising and have received much research interest. However, they are generally difficult to implement. Besides, unexpected termination of the computation process is often encountered, due to a highly distorted Lagrangian element or a very small Eulerian cell.

2.2.2. *Meshless methods*

Despite the great success of the grid-based methods in both CSM and CFD, their disadvantages have prompted researchers to look for new numerical methods. One important goal of the initial research is to modify the internal structure of the grid-based methods to become more adaptive and more robust. Additional effort was given to problems that cannot be easily treated by this class of methods, such as problems with moving boundaries (for Eulerian methods) and large deformation (for Lagrangian methods). Such an effort has given rise to the next generation of computational methods: the meshless methods. The main key of the meshless methods is to provide accurate and stable solutions for integral equations or partial differential equations (PDEs) with a set of arbitrary distributed nodes (or particles) without any connectivity between them. Contrary to the grid-based methods, the meshless methods normally do not require any predefined mesh to establish the system of algebraic equations for a given problem domain. Instead, only a set of nodes scattered within the problem domain and along the boundaries is used to represent the simulated geometry (Figure 2.4). In such a representation, no connectivity between nodes is required. The ideal requirements of the meshless methods are:

– non-necessity of a predefined mesh, at least in field variable interpolation;

– non-necessity of a predefined mesh at all throughout the computation process.

However, the meshless methods developed so far are not really ideal with regard to these two points. For example, some meshless methods require background cells (mesh) to derive the system of algebraic equations. The element free Galerkin (EFG) method [BEL 94] and the meshless local Petrov–Galerkin (MLPG) method [ATL 98] belong to this category. These two methods use background cells for integration of system matrices derived from the weak formulation.

Most of the existing meshless methods are based on the Lagrangian description. Therefore, construction of shape functions is necessary before or during the computation process. These shape functions will next be used to approximate the field variables at any point $\boldsymbol{x}$ within the problem domain as follows [2.2]:

$$\boldsymbol{u}(\boldsymbol{x}) = \sum_{i=0}^{n_T} \phi^i(\boldsymbol{x})\boldsymbol{u}_i \qquad [2.2]$$

where $\boldsymbol{u}$ is the field variable to be approximated (e.g. displacement), n_T is the total number of the nodes, $\boldsymbol{u}_i$ is the nodal field variable of the node i and ϕ^i is the shape function associated with the node i. According to the type of the shape functions, the meshless methods can be classified into two main categories: approximation methods and interpolation methods. The main features of each category are detailed hereafter.

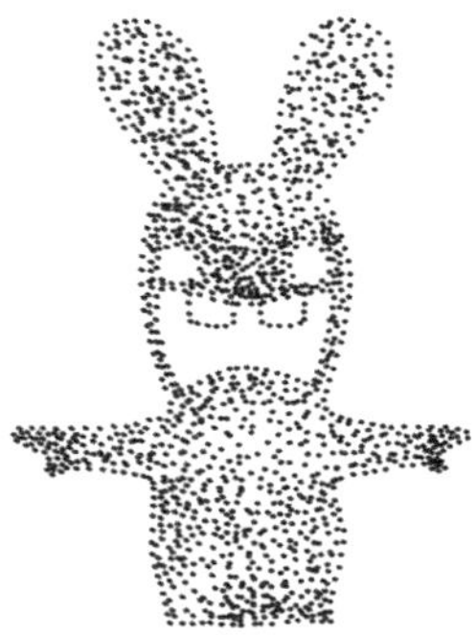

Figure 2.4. *Meshless discretization for rabbit simulation*

2.2.2.1. *Approximation methods*

Contrary to the Lagrangian grid-based methods, the shape functions in meshless methods are constructed for the nodes (not for element grids). In the case of approximation methods, these functions generally do not satisfy the *Kronecker delta function* property [2.3] at the nodes.

$$\phi^i(\boldsymbol{x}_j) = \begin{cases} 1 & \mathrm{i = j} \\ 0 & \mathrm{i \neq j} \end{cases} \qquad [2.3]$$

In other words, an approximated field variable based on these functions, called approximation shape functions, does not necessarily pass through the nodal values ($\boldsymbol{u}(\boldsymbol{x}_i) \neq \boldsymbol{u}_i$).

The first method developed in this category is the famous SPH [LIU 03, LIU 10, LUC 77, RAN 96]. This method was originally developed to model astrophysical phenomena, and later it has been widely extended for application to problems of solid and fluid mechanics [LUC 77]. The SPH method and its different variants, such as corrective smoothed particle method (CSPM) [LIU 03] and discontinuous smoothed particle hydrodynamics (DSPH) [LIU 03], are the major types of approximation methods. These methods have been incorporated into many commercial codes. Several other methods belonging to this category can be found in the literature. Among them, we can cite the generalized finite difference method (GFDM) [LIS 80], EFGM [BEL 94], reproducing kernel particle method (RKPM) [LI 02], etc. Several advantages have made the approximation methods very attractive:

– since no predefined mesh is required, approximation methods are able to deal with problems involving extremely large deformations;

– approximation methods are more adapted than the Eulerian grid-based methods to study problems with free surfaces, deformable boundary and moving interfaces;

– the approximation methods are more adapted than the grid-based methods to study fracture of materials. Indeed, they do not use any kind of mesh which can affect the cracking mechanisms;

– approximation methods are more adapted than grid-based methods to treat complex geometries which are represented only by a set of scattered nodes with no connectivity between them;

– applications of this class of methods are very wide, ranging from microscopic to macroscopic scales and even to astronomical scale, and from discrete systems to continuum systems;

– due to intensive research efforts, several approximation methods, such as SPH, are quickly approaching their mature stage;

– the accuracy of the solution is very weakly affected by the relative positions of the nodes. Therefore, it is very easy to add or remove nodes. This point is particularly attractive to define an adaptive refinement strategy for problems with localization.

However, the approximation methods have several drawbacks related to the imposition of boundary conditions and implementation requirements:

– to impose essential boundary conditions in a straightforward manner as in FEM, it is necessary that: (1) the approximated field variables pass through the nodal values (interpolation shape functions) and (2) the influence of the internal nodes vanishes at the border of the problem domain (linear shape functions at the border). Since, for approximation methods, generally none of these conditions is satisfied, it is not easy to impose essential boundary conditions. Therefore, special techniques are required to overcome this problem. In the literature, we can find several works on this subject and several techniques have been proposed. Among them, we can cite the Lagrange multiplier method [BEL 94], penalty method [ZHU 98] and the introduction of singular weight functions [KAL 97]. But, in most of the cases, these techniques are tedious and involve additional computational effort which can limit the application of these methods to problems with complex/moving interfaces and boundaries;

– the second drawback of the approximation methods is related to the numerical integration. In fact, almost all the shape functions used in meshless methods are rational (not polynomial). This makes the Gauss integration schemes not well adapted for this class of methods. Dolbow *et al.* [DOL 99] have demonstrated that, using these integration schemes, a great number of Gauss points are required which can induce an extensive additional cost of computation. Concerning the other popular integration technique, direct nodal integration, Chen *et al.* [CHE 01] have demonstrated that the application of this technique can involve numerical instability problems. To overcome these problems, several solutions have been proposed in the literature, such as the stabilized conforming nodal integration scheme [GON 04, CHE 01]. However, they are generally less accurate than the Gauss integration schemes for Lagrangian grid-based methods;

– another drawback of this class of methods is related to the definition of the influence area, called support domain or also influence domain, of the particles (nodes). In most of the cases, the support domain of a particle is taken as a sphere or parallelepiped centered on the particle. As discussed in several previous works [LIU 03, LIU 10], this support should contain a sufficient number of neighbor particles for computation stability and then should be sufficiently wide. However, a very wide support domain can affect the solution accuracy. The choice of the size of the support domain is not trivial and represents, at present, an open question [LEE 03];

– as reported in several previous works [BEL 96, LEE 03], the application of approximation methods to non-convex geometries, e.g. with the presence of concave boundaries, cavity or cracks, is very challenging. This is mainly due to difficulties in defining the support domain in the vicinity of the concave borders. In this case, the support domain can contain nodes which are not really neighbors, e.g. nodes on opposite sides of a crack;

– finally, compared to Lagrangian grid-based methods, approximation methods generally lead to poor and inaccurate solutions. The accuracy problem is much more amplified in the vicinity of the border of the problem domain.

Despite all the advantages of the approximation methods (meshless methods), the different drawbacks of these methods limit their application to several mechanical problems.

2.2.2.2. *Interpolation methods*

To overcome the limitations of the approximation methods, another generation of meshless methods has been developed: interpolation methods. In this generation, the field variables are approximated using interpolation shape functions, verifying the *Kronecker delta function* property [2.3] at the nodes. In other words, the approximated field variables pass through the nodal values ($\boldsymbol{u}(\boldsymbol{x}_i) = \boldsymbol{u}_i$). The first interpolation meshless method that has been developed is the natural element method (NEM) [BRA 95, TRA 94]. In this method, the interpolation shape functions are constructed based on the notion of natural neighbor (NN) interpolation. The NN interpolation is defined as a multivariate data interpolation scheme [SIB 81], which has been primarily used in data interpolation and modeling of geophysical phenomena. It relies on the concepts of Voronoï diagrams [VOR 07] and Delaunay tessellations [DEL 34], which are widely used in computational geometry, to construct the shape functions (called NN shape functions). With this type of interpolation, the NEM method has overcome most of the difficulties encountered with the earlier generation of meshless methods, such as definition of the support domains of the nodes. However, application of this method for non-convex geometries remains difficult. In effect, the NN shape functions are not strictly linear over non-convex boundaries, which makes difficult the imposition of Dirichlet boundary conditions on such boundaries. This is because NEM is closely related to the convex hull of a given set of points.

Recently, Cueto *et al.* [CUE 00, CUE 03] proposed an extension of NEM, called α-NEM, based on the concept of α-shapes or α-complexes, which are widely used in science visualization to give a shape to a cloud of points. The use of α-shapes in the context of NEM allows for construction of models entirely in terms of nodes and also ensures the linear precision of the shape functions even over non-convex boundaries. This is particularly true in the case of non-strongly non-convex domains. Indeed, as mentioned by Cueto *et al.* [CUE 00, CUE 03], the α-shapes used in such a method fail to ensure the linearity of the shape functions over highly non-convex boundaries (Figure 2.5), e.g. around a crack tip. α-NEM is, therefore, limited to non-strongly concave geometries. Otherwise, the same problem related to the imposition of the Dirichlet boundary conditions in the NEM is raised.

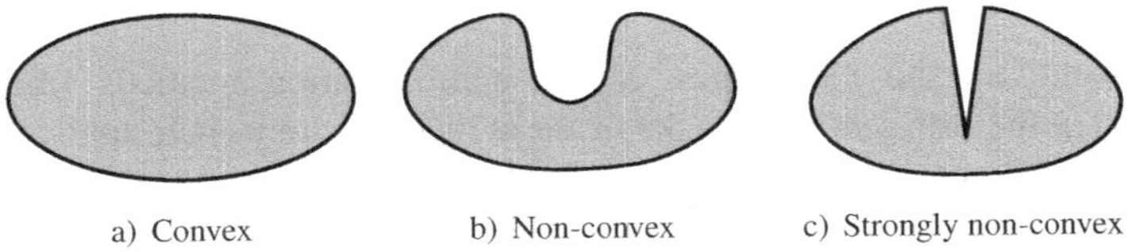

Figure 2.5. *Different geometries of problem domains*

More recently, Yvonnet *et al.* [YVO 05, YVO 04b] proposed another extension of NEM to ensure the linearity of the shape functions over all types of boundaries (even strongly non-convex): the constrained natural element method (CNEM). This method can be regarded as another extension of the NEM which retains its attractive properties and overcomes its difficulties to model strongly non-convex domains. The main originality of this approach lies in the introduction of a visibility criterion [YVO 05, YVO 04b] in the NEM to select the NNs of the nodes and to construct the interpolation. A modified so-called constrained Voronoï diagram is introduced to compute the shape functions. As a meshless method, CNEM has practically all the advantages of this class of methods. Also, with the introduction of the visibility criterion, it overcomes almost all its drawbacks.

2.3. Choice of continuum method

The grid-based methods have been widely applied in various areas of CSM and CFD, and are the dominant methods in computational mechanics. Table 2.1 presents a comparison between the different categories of the grid-based methods.

The Eulerian methods and the combined Lagrangian–Eulerian methods represent the "*weak links*" in the grid-based methods with respect to the objectives of this book. Indeed, the advantages of these methods are largely related to large and extremely large deformations. However, the discrete-continuum coupling method which will be developed in this work is mainly devoted to CSMs where deformations are not so significant. Besides, the DEM is based on Lagrangian description. Therefore, coupling this method with a CM based on the same description seems to be advantageous.

The Lagrangian methods which are mainly represented by the FEM are based on Lagrangian description and are more suited for CSM simulations.

Bearing in mind that the sought CM will be applied far from the regions of complex phenomena (which will be modeled by the DEM), the FEM represents a good candidate, among the grid-based methods, to be coupled with the DEM.

	Lagrangian methods	*Eulerian methods*	*Combined L-E*
Discretization	Mesh attached to material	Mesh fixed in the space	Updated mesh
Examples	FEM [ZIE 05a] X-FEM [MOË 99] BEM [HAL 94]	FVM [VER 07] FDM [SMI 85]	CEL [FEL 92] ALE [STE 04]
Track	Movement of material points	Mass, momentum and energy flux across mesh cells boundaries	Combined solution
Implementation	⋆⋆⋆	⋆⋆⋆	⋆
Time history at material points	⋆⋆⋆	⋆	⋆⋆
Boundaries and interfaces	⋆⋆⋆	⋆	⋆⋆
Complex geometries	⋆⋆⋆	⋆	⋆⋆
Large deformation	⋆⋆	⋆⋆⋆	⋆⋆
Discontinuities	⋆⋆	⋆	⋆
Computation robustness	⋆⋆⋆	⋆⋆	⋆⋆
Classification	1	3	2

⋆⋆⋆: Good/Easy ⋆⋆: Intermediate ⋆: Bad/Difficult

Table 2.1. *Comparison of the grid-based methods*

Concerning the meshless methods, the Lagrangian description is used for almost all the methods belonging to this class. Table 2.2 gives a comparison between the different categories of the meshless methods.

The main drawbacks of the approximation methods are related to the treatment of the problem domain boundaries, especially the imposition of essential boundary conditions (e.g. displacements). The choice of an approximation method as a part of the coupling approach can then entail more difficulties in the coupling implementation. Indeed, to ensure correct communication between the coupled methods, it is important to correctly treat the interface between them. Therefore, additional effort must be undertaken when approximation method is used. For this reason, this class of methods will not be retained.

Subclasses	*Approximation methods*	*Interpolation methods*
Discretization	Scattered nodes	Scattered nodes
Examples	SPH [LUC 77, LIU 03] PUFEM [MEL 96] EFGM [BEL 94] MLPG [ATL 98] RKPM [LI 02] PIM [LIU 01] FPM [OÑA 96, OÑA 98, OÑA 00]	NEM α-NEM CNEM
Track	Movement of material points	Movement of material points
Implementation	⋆⋆⋆	⋆⋆
Time history at material points	⋆⋆	⋆⋆⋆
Boundaries and interfaces	⋆⋆	⋆⋆⋆
Complex geometries	⋆⋆	⋆⋆⋆
Large deformation	⋆⋆⋆	⋆⋆⋆
Discontinuities	⋆⋆	⋆⋆
Computation robustness	⋆⋆	⋆⋆⋆
Classification	2	1

⋆⋆⋆: Good/Easy ⋆⋆: Intermediate ⋆: Bad/Difficult

Table 2.2. *Comparison of meshless methods*

The interpolation methods have overcome almost all the difficulties encountered with the approximation methods. The first method in this class which is the NEM is mainly adapted for convex domains. Moreover, the α-NEM, which is an extension of the NEM, is more suited for non-extremely non-convex domains. With the aim of developing a generic coupling method able to deal with all types of domains (even extremely non-convex), these two methods will not be retained. The CNEM which is also an extension of the NEM has circumvented the geometry limitations, while keeping the attractive properties of the factory method (NEM). Therefore, it is a good candidate to be coupled with the DEM.

At this stage, two numerical methods present good candidates to be coupled with DEM: a grid-based method (FEM) and a meshless method (CNEM). To select the most appropriate method, a comparison between them on some fundamental aspects is given in Table 2.3. These two methods are very related through the following points: (1) the need of an underlying structure; (2) the boundary conditions can be directly applied, due to the interpolation character and the linearity of the shape functions at the boundaries and interfaces. Applied on the same mechanical problem, FEM is generally less expensive in terms of Central Processing Unit (CPU) time and more robust than CNEM [SUK 98]. Indeed, a step for construction of Delaunay tessellations and the computation of shape functions is needed in the case of CNEM which can be costly. Moreover, the use of rational shape functions in CNEM (vs. polynomial shape functions in FEM) makes the CNEM integration more laborious and less robust. However, since only scattered nodes (without any predefined connectivity between them) are used in CNEM for spatial discretization, this method is more suited for complex geometries. Also, CNEM is more advantageous to treat relatively large deformations characterizing the response of solids in highly dynamics, which is the scope of this work. Another point fostering the choice of CNEM, compared to FEM, is due to its "meshless" nature. Since, DEM is also a meshless method, the coupling between CNEM and DEM seems more natural. No additional effort, due to the presence of mesh at the interface between these methods, is required. For these reasons, CNEM is finally retained to be coupled with DEM in this work. The main specificities of this method will be reviewed in the next section.

2.4. The constrained natural element method

As an NEM extension, the CNEM is very similar to the factory method (NEM) and keeps all its attractive properties. The main difference between them lies in the introduction of a visibility criterion restricting the NNs to correctly model strongly non-convex problems. For clarity, the main features of the factory method (NEM) will first be reviewed. Then, the visibility criterion as well as the resulting modifications will be introduced.

2.4.1. *Natural neighbor interpolation*

The notion of NNs and NN interpolation was introduced by Sibson [SIB 80] for data fitting and smoothing of highly irregular set of nodes. The NN interpolation is based on the Voronoï diagram and the Delaunay tessellations which were originally proposed by mathematicians [DEL 34, VOR 07] and later applied in several engineering fields.

	FEM	*CNEM*
Discretization	⋆⋆	⋆⋆⋆
Implementation	⋆⋆⋆	⋆⋆
Time history at material points	⋆⋆⋆	⋆⋆⋆
Boundaries and interfaces	⋆⋆⋆	⋆⋆⋆
Complex geometries	⋆⋆	⋆⋆⋆
Large deformation	⋆	⋆⋆
Discontinuities	⋆	⋆⋆
Computation robustness	⋆⋆⋆	⋆⋆
Classification	2	1
⋆⋆⋆: Good	⋆⋆: Intermediate	⋆: Bad

Table 2.3. *Comparison between FEM and CNEM*

2.4.1.1. *Voronoï diagram*

The Voronoï diagram is defined as a set of cells associated with the scattered nodes such that all the points in a particular cell T_i are closer to the corresponding node i than to any other node (Figure 2.6). The Voronoï diagram of a given set of nodes is unique, it fills the convex hull of these nodes. Mathematically, this diagram is defined in any dimension d as follows:

$$T_i = \left\{ \boldsymbol{x} \in \mathbb{R}^d, \; \|\boldsymbol{x}\boldsymbol{x}_i\| < \|\boldsymbol{x}\boldsymbol{x}_j\|, \; \forall i \neq j \right\}, \qquad i = 1..n_T \qquad [2.4]$$

where n_T is the total number of nodes and $\|\boldsymbol{x}\boldsymbol{x}_i\|$ represents the distance between a point of $\boldsymbol{x}$ coordinates and the node i located in $\boldsymbol{x}_i$.

In duality with the Voronoï diagram, the concept of the Delaunay tessellation (or triangulation in two-dimensional (2D)) can be introduced. This concept was proposed by Voronoï [VOR 07] and extended by Delaunay [DEL 34]. It consists of connecting the nodes whose the Voronoï cells share some boundaries to construct the Delaunay tessellations (triangles in 2D and tetrahedrons in three-dimensional (3D)). An important property of the Delaunay tessellations is that the circumscribed spheres (mathematical definition of a sphere is used here, this is circle in 2D or sphere in 3D) of these elements (Delaunay tessellations) do not contain any node. Figure 2.6 presents an example of Voronoï diagram and the associated Delaunay tessellation in 2D.

The NNs of a node i are those connected to i by an edge in the Delaunay tessellations. In other words, nodes having Voronoï cells that share some boundaries are NNs. If the nodes are relatively spaced in some regions or distributed in highly anisotropic manner, the set of NNs will also reflect these features. However, it always represents the best set of surrounding neighbors. The NNs are natural candidates to define a basis for construction of the shape functions and to define a local interpolation scheme.

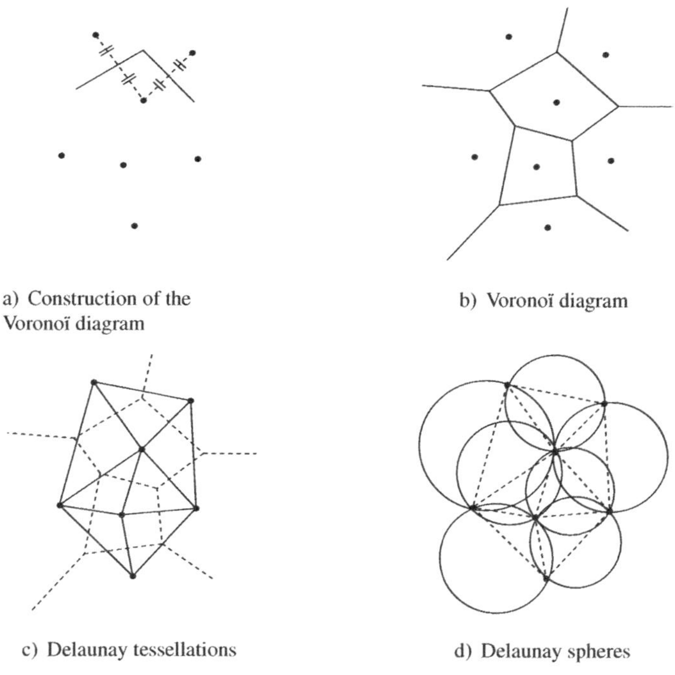

Figure 2.6. *Geometric construction of the Voronoï diagram and the associated Delaunay tesselation for a set of nodes in the plane (2D)*

2.4.1.2. *NN shape functions*

To define the neighbor relation of any point placed into the Delaunay tessellation, Sibson [SIB 80] used the concept of second-order Voronoï diagram which is defined as a set of cells T_{ij}. A cell T_{ij} is the locus of the points $\boldsymbol{x}$ having i as the closest node and j as the second closest node. Mathematically, this diagram is defined in any dimension d as follows:

$$T_{ij} = \left\{ \boldsymbol{x} \in \mathbb{R}^d, \; \|\boldsymbol{x}\boldsymbol{x}_i\| < \|\boldsymbol{x}\boldsymbol{x}_j\| < \|\boldsymbol{x}\boldsymbol{x}_k\|, \; \forall i \neq j \neq k \right\}, \quad i = 1..n_T \qquad [2.5]$$

where n_T is the total number of nodes and $\|\boldsymbol{x}\boldsymbol{x}_i\|$ represents the distance between a point of $\boldsymbol{x}$ coordinates and the node i located in $\boldsymbol{x}_i$. Several types of shape functions are used in the NN interpolation.

2.4.1.2.1. Sibson shape functions

Assuming that $\mathcal{K}$ denotes the Lesbegue measure (length, area or volume in, respectively, 1D, 2D or 3D), $\mathcal{K}_x$ denotes the Lesbegue measure of T_x and $\mathcal{K}_{xi}$ denotes the Lesbegue measure of T_{xi}. The Sibson shape function ϕ^i associated with a node i is defined as the ratio of $\mathcal{K}_{xi}$ and $\mathcal{K}_x$, where $\boldsymbol{x}$ is the coordinates vector of the point at which the shape function is calculated.

$$\phi^i(\boldsymbol{x}) = \frac{\mathcal{K}_{xi}}{\mathcal{K}_x} \qquad [2.6]$$

In [2.6], $\mathcal{K}_x$ can also be calculated as $\mathcal{K}_x = \sum_{i=1}^{n_N} \mathcal{K}_{xi}$, where n_N is the number of the NNs of $\boldsymbol{x}$. Figure 2.7(a) gives an illustration of 2D calculation of Sibson shape functions. In 2D, the Sibson shape function ϕ^i associated with a node i is calculated in $\boldsymbol{x}$ as the ratio of the A and B polygon areas (Figure 2.7):

$$\phi^i(\boldsymbol{x}) = \frac{area_A}{area_B} \qquad [2.7]$$

The Sibson shape functions have C^0 continuity at the nodes, C^1 continuity at the Delaunay spheres (circles in 2D) and C^∞ continuity anywhere else.

2.4.1.2.2. Laplace (non-sibsonian) shape functions

Other NN shape functions, called Laplace or non-sibsonian shape functions, are used in the NN interpolation [BEL 97]. Assuming that f_{xi} is the Lesbegue measure of the Voronoï facet (length in 2D or area in 3D) separating the point of $\boldsymbol{x}$ coordinates and the node i, and d_{xi} is the distance between them. The Laplace shape function associated with i is calculated in $\boldsymbol{x}$ as:

$$\phi^i(\boldsymbol{x}) = \frac{\alpha_{xi}}{\sum_{i=1}^{n_N} \alpha_{xj}}, \quad \text{with } \alpha_{xi} = \frac{f_{xi}}{d_{xi}} \qquad [2.8]$$

where n_N is the number of the NNs of $\boldsymbol{x}$. Figure 2.7(b) gives an illustration of 2D calculation of Laplace shape functions. The Laplace shape functions have C^0 continuity at the nodes and at the Delaunay spheres (circles in 2D), and C^∞ continuity anywhere else. These functions are less expensive in terms of CPU than the Sibson shape functions. Only areas (lengths in 2D) require to be evaluated in these shape functions, whereas, volumes (area in 2D) must be calculated in the Sibson shape functions. However, the Sibson shape functions are more regular at the Delaunay spheres (circles in 2D) (C^1 continuity).

2.4.1.2.3. Other NN shape functions

A generalization of NN-based shape functions, called $\mathrm{k^{th}}$ – order standard coordinates, has been proposed by Hiyoshi and Sugihara [HIY 02]. These functions have C^0 continuity at the nodes, C^k continuity (k is a non-negative integer) at the Delaunay spheres and C^∞ continuity anywhere else. The cases corresponding to $k = 0$ and $k = 1$ coincide, respectively, with the Laplace and Sibson shape functions. It should be noted that the computation time of the shape functions is proportional to the continuity order k at the Delaunay spheres. Therefore, using highly regularized shape functions (large k) is very time-consuming and can become crippling.

NOTE.– In this work, Sibson shape functions are retained for interpolation. These functions are more regular than the Laplace shape functions. Also, they are not very costly in terms of CPU compared to the $\mathrm{k^{th}}$ – order standard coordinates with ($k > 1$). Using Sibson shape functions, the NN interpolation of a vector-valued function $\boldsymbol{u}$ can be derived:

$$\boldsymbol{u}^h(\boldsymbol{x}) = \sum_{i=1}^{n_N} \phi^i(\boldsymbol{x})\, \boldsymbol{u}_i \qquad [2.9]$$

where n_N is the number of NN nodes of the point having $\boldsymbol{x}$ coordinates, ϕ^i is the Sibson shape function associated with a node i and $\boldsymbol{u}_i$ is the nodal value of $\boldsymbol{u}$ at the node i of $\boldsymbol{x}_i$ coordinates ($\boldsymbol{u}_i = \boldsymbol{u}(\boldsymbol{x}_i)$).

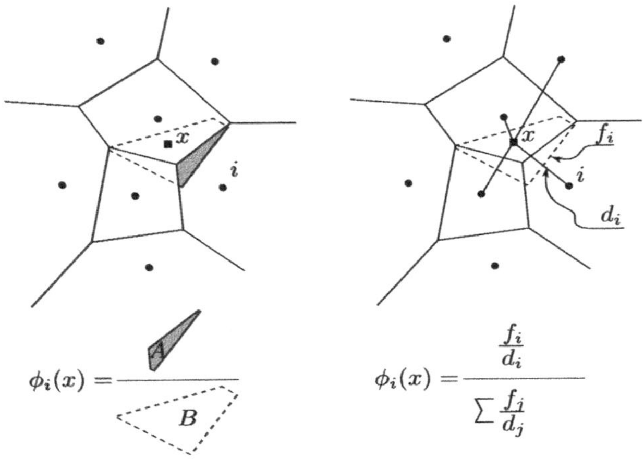

Figure 2.7. *Calculation of Sibson shape functions a) and Laplace shape functions b)*

2.4.1.3. *Support of NN shape functions*

The support of an NN shape function $\phi^i(\boldsymbol{x})$ associated with i is defined as the space occupied by the union of the n_N circumspheres (circumcircles in 2D) of the Delaunay tetrahedrons (triangles in 2D) in which i is a vertex; n_N is the number of the NNs of i. Figure 2.8 presents an example of a 2D NN shape function support. Unlike in other meshfree methods, the support domains of the NN shape functions are automatically defined by the NNs. These supports are always represented by the best set of surrounding neighbors even in the case of irregular of highly anisotropic distribution of nodes.

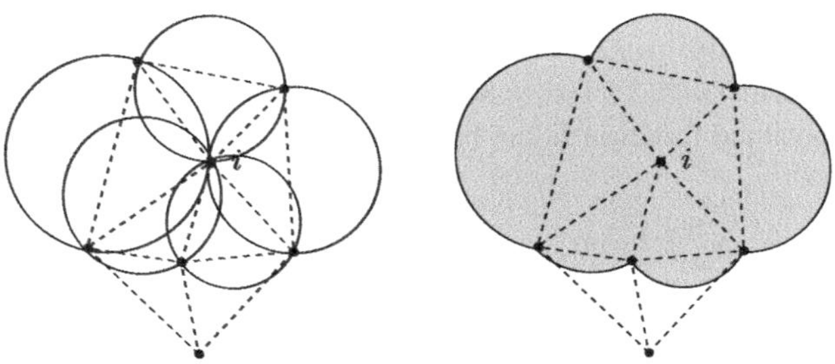

Figure 2.8. *Support of NN shape function associated with a node i (2D)*

2.4.1.4. *Properties of NN shape functions*

The main properties of the NN shape functions are:

– *The Kronecker delta property*: this property states that the NN interpolation passes through the nodal values.

$$\phi^i(\boldsymbol{x}_j) = \begin{cases} 1 & \mathrm{i} = \mathrm{j} \\ 0 & \mathrm{i} \neq \mathrm{j} \end{cases} \qquad [2.10]$$

– *Compact support*: the NN shape functions are compactly supported. This means that they are non-zero only in delimited regions and vanish beyond. The NN interpolation is, therefore, transformed from global operation to local operation. This leads to a set of sparse discretized system matrices, which considerably reduces the computational effort.

– *Partition of unity*: by construction, the NN shape functions verify the following relation:

$$\sum_{i=1}^{n_T} \phi^i(\boldsymbol{x}) = 1 \quad \forall \boldsymbol{x} \in \Omega \qquad [2.11]$$

where n_T is the total number of nodes. This property is also called 0th – order (C^0) consistency and means that the NN interpolation can exactly reproduce constant functions.

– *Linear consistency*: the NN shape functions also satisfy the local coordinate property (1th – order (C^1) consistency) defined as:

$$\boldsymbol{x} = \sum_{i=1}^{n_T} \phi^i(\boldsymbol{x})\boldsymbol{x}_i \quad \forall \boldsymbol{x} \in \Omega \qquad [2.12]$$

This property combined with the partition of unity property implies that the NN interpolation has the linear completeness property. The NN interpolation exactly reproduces any linear polynomial. It should be noted that, in the resolution of PDEs based on a weak formulation (e.g. the Galerkin formulation), there is a minimum consistency requirement to ensure the convergence of the numerical results, depending on the PDEs order. For the Galerkin formulation, C^k consistency is required to solve PDEs of $2k$ order.

– *Linear variation on convex boundaries*: the NN shape functions are strictly linear on convex boundaries. Combined with the *Kronecker delta property*, this property implies that the essential boundary conditions can directly be applied. However, in the case of non-convex boundaries, this property is not satisfied. This problem is circumvented by introducing the notion of visibility criterion and then the constrained natural neighbor (CNN) interpolation.

2.4.2. *Visibility criterion*

This criterion states that a point of $\boldsymbol{x}$ coordinates is in the influence domain of a node i if $\boldsymbol{x}$ is within the region where the shape function ϕ^i associated with i is non-zero ($\phi^i(\boldsymbol{x}) \neq 0$) and it is visible from the node i when the boundaries are assumed to be opaque. Assuming that Ω_C is the whole continuum domain and $\Omega_i = \{\boldsymbol{x}|\boldsymbol{x} \in \Omega_C,\ \phi^i(\boldsymbol{x}) \neq 0\}$ is the original (convex) influence domain of a node i, the visibility criterion is defined mathematically as:

$$\Omega_i^{visible} = \{\boldsymbol{x}|\boldsymbol{x} \in \Omega_i,\ l_{x \to x_i} \cap \partial\Omega_i = \varnothing\} \qquad [2.13]$$

where $l_{x \to x_i}$ represents the straight line relying $\boldsymbol{x}$ and $\boldsymbol{x}_i$ (coordinates of i) and $\partial\Omega_i$ is the boundary of Ω_i.

2.4.3. *Constrained natural neighbor interpolation*

Taking into account the visibility criterion, the Voronoï diagram is extended to become the constrained Voronoï diagram [EDE 01] which is defined as:

$$T_i^C = \{\boldsymbol{x} \in \mathbb{R}^d,\ \|\boldsymbol{x}\boldsymbol{x}_i\| < \|\boldsymbol{x}\boldsymbol{x}_j\|,$$
$$\forall i \neq j,\ l_{x \to x_i} \cap \partial\Omega_i = \varnothing,\ l_{x \to x_j} \cap \partial\Omega_j = \varnothing\},\ i = 1..n_T \qquad [2.14]$$

Concerning the constrained Delaunay tessellation, it is always possible to construct the constrained Delaunay triangulation for a given set of nodes in the plane (in $2D$). However, as reported by several researchers [SHE 98, YVO 04a], its generalization to $3D$ does not always exist without adding new nodes. Some techniques have been proposed in the literature to construct the constrained Delaunay tessellations in $3D$ by adding stainer points [SHE 98].

With the introduction of the visibility criterion, the NNs of any point are confined to the constrained NNs which are visible by the considered point. Once the constrained Voronoï diagram as well as the associated Delaunay tessellation are constructed, the selection of the constrained NNs can directly be determined, following the same procedure as in NEM. The associated CNN shape functions ϕ^{i^C} can also be determined using the constrained

Voronoï diagram and the constrained second-order Voronoï diagram. In such diagrams, any Voronoï cell facet between two non-mutually-visible nodes is removed. In this case, the CNN shape functions ϕ^{i^C} restore the linearity property on any type of boundaries. The NN interpolation given by [2.9] becomes:

$$\boldsymbol{u}^h(\boldsymbol{x}) = \sum_{i=1}^{n_N^C} \phi^{i^C}(\boldsymbol{x})\, \boldsymbol{u}_i \qquad [2.15]$$

where n_N^C is the number of the CNN nodes.

2.4.4. *Numerical integration*

In computational mechanics, the global weak formulation of the studied problem can be derived using several approaches, such as the Galerkin approach used in CNEM. This formulation is given in continuous form which is very difficult, if not possible, to be analytically solved. However, approximate solution can be obtained numerically by changing the continuous weak formulation into a discrete system of equations. The integrals in this system are evaluated numerically using several techniques. Among them, we can cite the Gauss integration and the direct nodal integration. In the case of CNEM, the Gauss technique may be inaccurate since the shape functions are not generally polynomials and the supports of these functions do not coincide with the integration cells [CHE 01, GON 04]. Also, the direct nodal integration (using the nodes as integration points) typically leads to numerical instabilities due to underintegration of the weak formulation, as reported in the literature [BEI 96]. Recently, new numerical integration procedures have been proposed in the framework of meshless methods. The major of them is the stabilized conforming nodal integration technique (SCNI) developed by Chen *et al.* [CHE 01]. This technique, which was originally developed for strain localization problems, relies on the notion of gradient smoothing to stabilize the nodal integration as follows:

$$\tilde{\nabla}\boldsymbol{u}^h(\boldsymbol{x}_i) = \int_{\Omega_i} w(\boldsymbol{x}, \boldsymbol{x} - \boldsymbol{x}_i)\, \nabla \boldsymbol{u}^h(\boldsymbol{x}_i)\, d\Omega \qquad [2.16]$$

where $\boldsymbol{x}_i$ are the coordinates of the node i, Ω_i is a subdomain surrounding i (generally taken as the Voronoï cell associated with i). w is a weight function having a compact support (vanishes beyond a certain region) and satisfying these properties: $w \geq 0$ and $\int_{\Omega_i} w(\boldsymbol{x}, \boldsymbol{x} - \boldsymbol{x}_i)\, d\Omega = 1$. This function is generally taken as the step function defined on Ω_i, supposing that V_i is the Lesbegue measure of Ω_i:

$$w(\boldsymbol{x}, \boldsymbol{x} - \boldsymbol{x}_i) = \begin{cases} \frac{1}{V_i} \text{ if } \boldsymbol{x} \in \Omega_i \\ 0 \text{ if } \boldsymbol{x} \notin \Omega_i \end{cases} \quad [2.17]$$

Using [2.16] and [2.17], the smoothed strain at the node i can be determined as:

$$\tilde{\varepsilon}^h_{kl}(\boldsymbol{x}_i) = \frac{1}{V_i} \int_{\Omega_i} \varepsilon^h_{kl}(\boldsymbol{x})\, d\Omega \quad [2.18]$$

Applying the divergence theorem, the integral [2.18] can be transformed to hyperplane contour integral (the hyperplane is defined here as a subset of dimension $d-1$, d is the dimension of Ω_i):

$$\tilde{\varepsilon}^h_{kl}(\boldsymbol{x}_i) = \frac{1}{2V_i} \int_{\partial\Omega_i} (u^h_k n_l + u^h_l n_k)\, d\Gamma \quad [2.19]$$

$\boldsymbol{n}$ is the outward-pointing unit normal vector to $\partial\Omega_i$. The contour integrals can then be approximated by successive application of the Gauss integration technique to each facet of Ω_i. Introducing now the CNN interpolation, i.e. replacing $\boldsymbol{u}^h$ by its expression given by [2.15], the strain tensor at i can be written in engineering notation as:

$$\tilde{\boldsymbol{\varepsilon}}^h(\boldsymbol{x}_i) = \sum_{l=1}^{n_N^C} [B_l(\boldsymbol{x}_i)]\, \{\boldsymbol{u}_l\} \quad [2.20]$$

with: $\tilde{\boldsymbol{\varepsilon}}^h = {}^t\left\{ \tilde{\varepsilon}^h_{11}\; \tilde{\varepsilon}^h_{22}\; \tilde{\varepsilon}^h_{33}\; 2\,\tilde{\varepsilon}^h_{23}\; 2\,\tilde{\varepsilon}^h_{13}\; 2\,\tilde{\varepsilon}^h_{12} \right\}$, $\{\boldsymbol{u}_l\} = {}^t\left\{ u_{l1}\; u_{l2}\; u_{l3} \right\}$, n_N^C is the number of the CNN nodes (visible neighbors of the node i) and $[B_l]$ is the smoothed gradient matrix [CHE 01, GON 04] associated with the node l:

$$[B_l(\boldsymbol{x}_i)] = \begin{bmatrix} b_{l1}(\boldsymbol{x}_i) & 0 & 0 \\ 0 & b_{l2}(\boldsymbol{x}_i) & 0 \\ 0 & 0 & b_{l3}(\boldsymbol{x}_i) \\ 0 & b_{l3}(\boldsymbol{x}_i) & b_{l2}(\boldsymbol{x}_i) \\ b_{l3}(\boldsymbol{x}_i) & 0 & b_{l1}(\boldsymbol{x}_i) \\ b_{l2}(\boldsymbol{x}_i) & b_{l1}(\boldsymbol{x}_i) & 0 \end{bmatrix},$$

$$b_{lk}(\boldsymbol{x}_i) = \frac{1}{V_i} \int_{\partial\Omega_i} \phi^{l^C}(\boldsymbol{x})\, n_k\, d\Gamma \quad [2.21]$$

2.5. Conclusion

The great scientific interest in computational mechanics has yielded numerous CMs that can be used to model mechanical problems at the engineering scale. The present chapter focused on the choice of the appropriate CM to be coupled with DEM for highly dynamic studies. To simplify this task, the CMs were divided into two classes: grid-based methods and meshless methods. Comparison of these classes with respect to the expectations of this work allowed for retaining the class of meshless methods. This class, in turn, can be divided into approximation and interpolation methods. From these categories, the interpolation methods offer better solutions for difficulties related to imposition of boundary conditions. Therefore, they are best adapted to be coupled with DEM. Among the interpolation meshless methods, the CNEM is the most recent and includes practically all the advantages of this category. Therefore, it was retained to be part of the discrete-continuum coupling. The main features of this method were then recalled to better understand the development of the coupling formulation, which is the subject of Chapter 3.

3

Development of Discrete-Continuum Coupling Method Between DEM and CNEM

3.1. Introduction

The most interesting problems in fast dynamics are often also the most difficult to study. These problems generally involve a multitude of complex multi-scale phenomena which cannot be easily treated by traditional continuum methods (CMs). The discrete element method (DEM) proposed in the first book of this set [JEB 15] offers an alternative approach to investigating such problems. The major drawback of this method is that it is very demanding in terms of computation time, and can only be applied for small physical systems. However, in most of the cases of highly dynamic problems, the regions requiring fine-scale analysis by DEM are small with respect to the entire studied domain. Therefore, the development of a numerical tool combining the DEM accuracy and the CMs fastness is very appealing to study these problems. In this context, the present book proposes to elaborate a discrete-continuum coupling approach, adapted for highly dynamic problems. To achieve this aim, the previous chapters have dealt with the choice of the appropriate CM as well as the coupling technique that will be used to develop the discrete-continuum coupling approach. Among the CMs, the constrained natural element method (CNEM) was selected. To couple this method with DEM, the bridging domain (BD) technique was retained. This technique is well adapted for highly dynamic multi-scale problems. Indeed, it allows the high-frequency waves (HFWs) that are not supported by CNEM to be naturally canceled by the projection mechanism in the bridging region, provided that the coupling parameters are well set out.

Using these choices, the present chapter aims to put into practice the development of the proposed coupling approach. First, it will be detailed how the DEM-CNEM coupling approach is performed. Then, the different coupling parameters involved in this approach will be studied to simplify their setting in practice and then the application of the developed coupling approach on complex problems. Based on this study, several recommendations and trends simplifying the choice of these parameters will be drawn. Applying these recommendations, the DEM-CNEM coupling approach will be validated using several dynamic reference tests: dynamic tension, bending, torsion and impact tests using three-dimensional (3D) beams.

3.2. Discrete-continuum coupling method: DEM-CNEM

The present section details the different steps followed to develop the DEM-CNEM coupling approach. Since this approach is dedicated to highly dynamic problems, velocity constraints are chosen to enforce the DEM-CNEM compatibility in the bridging region. This allows kinetic energy to be better exchanged between DEM and CNEM. To simplify the application of these constraints, the variational approach of virtual power will be used to derive the coupling formulation. This explicitly shows velocity terms that will be controlled in the bridging region.

3.2.1. *DEM-CNEM coupling formulation*

A domain Ω_G is considered with boundary $\partial\Omega = \partial\Omega^u \cup \partial\Omega^t$ ($\partial\Omega^u \cap \partial\Omega^t = \phi$) such that essential ($\boldsymbol{u}_g$) and natural ($\boldsymbol{t}_g$) boundary conditions are, respectively, prescribed on $\partial\Omega^u$ and $\partial\Omega^t$. This domain is divided into two subdomains Ω_D and Ω_C with a bridging region. These subdomains are, respectively, modeled using the discrete (DEM) and the continuum (CNEM) method (Figure 3.1). An isotropic linear elastic behavior and small strain gradients are assumed for simplicity. For the sake of clarity, the governing equations of both Ω_D and Ω_C will first be derived without taking into account the coupling conditions section 3.2.1.1 and section 3.2.1.2. These conditions will be introduced later in section 3.2.1.3.

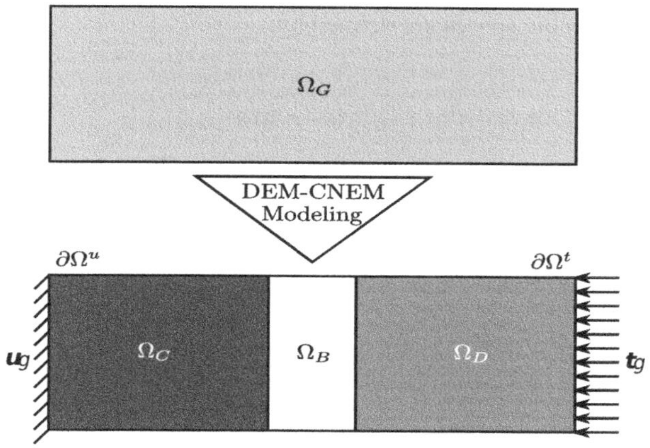

Figure 3.1. *Discrete-continuum decomposition*

3.2.1.1. *DEM formulation*

In an isolated system of the discrete subdomain Ω_D which is a set of n_D spherical discrete elements interacting via elastic cohesive beams, the governing equations can be written as: find $\left(\boldsymbol{d}_i, \boldsymbol{\theta}_i, \boldsymbol{f}_i^{int}, \boldsymbol{c}_i^{int}\right) \in \mathbb{R}^3 \times \mathbb{R}^3 \times \mathbb{R}^3 \times \mathbb{R}^3$ for $i = 1..n_D$ and $t \in [0, t_{end}]$ such that, given the initial conditions:

$$\begin{cases} \boldsymbol{f}_i^{ext} + \boldsymbol{f}_i^{int} = m_i \ddot{\boldsymbol{d}}_i \\ \boldsymbol{c}_i^{ext} + \boldsymbol{c}_i^{int} = I_i \ddot{\boldsymbol{\theta}}_i \end{cases} \qquad [3.1]$$

where $\boldsymbol{d}_i$, $\boldsymbol{\theta}_i$, m_i and I_i represent, respectively, the displacement vector, rotation vector, mass and mass moment of inertia of the particle i. $\boldsymbol{f}_i^{int}$ and $\boldsymbol{c}_i^{int}$ are, respectively, the internal force and torque applied on the particle i by its neighbors via the cohesive links. $\boldsymbol{f}_i^{ext}$ and $\boldsymbol{c}_i^{ext}$ are, respectively, the external force and torque applied on the particle i. In keeping with the weak formulation in the continuum methods, equation [3.1] can be reformulated as follows: find $(\boldsymbol{d}, \boldsymbol{\theta}, \boldsymbol{f}, \boldsymbol{c}) \in \mathcal{D}^{ad} \times \mathcal{O}^{ad} \times \mathcal{F}^{ad} \times \mathcal{C}^{ad}$ such that $\forall \left(\delta\dot{\boldsymbol{d}}, \delta\dot{\boldsymbol{\theta}}\right) \in \dot{\mathcal{D}}^{ad,0} \times \dot{\mathcal{O}}^{ad,0}$, given the initial conditions:

$$\sum_{i=1}^{n_D} \boldsymbol{f}_i^{ext}.\delta\dot{\boldsymbol{d}}_i + \sum_{i=1}^{n_D} \boldsymbol{f}_i^{int}.\delta\dot{\boldsymbol{d}}_i + \sum_{i=1}^{n_D} \boldsymbol{c}_i^{ext}.\delta\dot{\boldsymbol{\theta}}_i + \sum_{i=1}^{n_D} \boldsymbol{c}_i^{int}.\delta\dot{\boldsymbol{\theta}}_i$$
$$= \sum_{i=1}^{n_D} m_i \ddot{\boldsymbol{d}}_i.\delta\dot{\boldsymbol{d}}_i + \sum_{i=1}^{n_D} I_i \ddot{\boldsymbol{\theta}}_i.\delta\dot{\boldsymbol{\theta}}_i \qquad [3.2]$$

where the admissible spaces are defined by:

– $\mathcal{D}^{ad} = \{\boldsymbol{d} = \boldsymbol{d}_i(t),\ i = 1..n_D\ \forall\, t \in [0, t_{end}]\}$

– $\mathcal{O}^{ad} = \{\boldsymbol{\theta} = \boldsymbol{\theta}_i(t),\ i = 1..n_D\ \forall\, t \in [0, t_{end}]\}$

– $\mathcal{F}^{ad} = \left\{\boldsymbol{f} = \boldsymbol{f}_i^{int}(t),\ i = 1..n_D\ \forall\, t \in [0, t_{end}]\right\}$

– $\mathcal{C}^{ad} = \left\{\boldsymbol{c} = \boldsymbol{c}_i^{int}(t),\ i = 1..n_D\ \forall\, t \in [0, t_{end}]\right\}$

– $\dot{\mathcal{D}}^{ad,0} = \left\{\dot{\boldsymbol{d}} = \dot{\boldsymbol{d}}_i(t),\ i = 1..n_D;\ \dot{\boldsymbol{d}} = \boldsymbol{0} \text{ on } \partial\Omega_D^u;\ \forall\, t \in [0, t_{end}]\right\}$

– $\dot{\mathcal{O}}^{ad,0} = \left\{\dot{\boldsymbol{\theta}} = \dot{\boldsymbol{\theta}}_i(t),\ i = 1..n_D;\ \dot{\boldsymbol{\theta}} = \boldsymbol{0} \text{ on } \partial\Omega_D^u;\ \forall\, t \in [0, t_{end}]\right\}$

3.2.1.2. *CNEM formulation*

As an isolated system, the governing equations in the continuum subdomain Ω_C can be written as: given the initial conditions, find $(\boldsymbol{u}, \boldsymbol{\sigma}) \in \left[\mathcal{H}^1(\Omega_C)\right]^3 \times \left[\mathcal{L}^2(\Omega_C)\right]^6$ for $\boldsymbol{x} \in \Omega_C(t)$ and $t \in [0, t_{end}]$ such that:

$$\begin{cases} div(\boldsymbol{\sigma}) + \rho \boldsymbol{f} = \rho \ddot{\boldsymbol{u}} & \text{in } \Omega_C \\ \boldsymbol{\sigma} = \boldsymbol{A} : \varepsilon(\boldsymbol{u}) & \\ \varepsilon(\boldsymbol{u}) = \frac{1}{2}(\nabla \boldsymbol{u} + \nabla^t \boldsymbol{u}) & \\ \boldsymbol{u} = \boldsymbol{u}_g & \text{on } \partial\Omega_C^u \\ \boldsymbol{\sigma}.\boldsymbol{n} = \boldsymbol{t}_g & \text{on } \partial\Omega_C^t \end{cases} \qquad [3.3]$$

where ρ is the density, $\boldsymbol{u}$ is the continuum displacement vector, $\boldsymbol{\sigma}$ is the Cauchy stress tensor, ε is the stain tensor, $\boldsymbol{A}$ is the stiffness tensor, $\boldsymbol{f}$ is the body force vector, $\boldsymbol{u}_g$ and $\boldsymbol{t}_g$ are, respectively, the given (prescribed) displacement and traction vectors on $\partial\Omega_C^u$ and $\partial\Omega_C^t$ and $\boldsymbol{n}$ is the outward-pointing unit normal vector. The continuum weak formulation associated with equation [3.3] can be written as follows: given the initial conditions, find $\boldsymbol{u} \in \mathcal{U}^{ad}$ such that $\forall\, \delta\dot{\boldsymbol{u}} \in \dot{\mathcal{U}}^{ad,0}$:

$$\int_{\partial\Omega_C^T} \delta\dot{\boldsymbol{u}} \cdot \boldsymbol{t}_g \, d\Gamma - \int_{\Omega_C} \varepsilon(\delta\dot{\boldsymbol{u}}) : \boldsymbol{A} : \varepsilon(\boldsymbol{u}) \, d\Omega + \int_{\Omega_C} \rho \, \delta\dot{\boldsymbol{u}} \cdot \boldsymbol{f} \, d\Omega = \int_{\Omega_C} \rho \, \delta\dot{\boldsymbol{u}} \cdot \ddot{\boldsymbol{u}} \, d\Omega \qquad [3.4]$$

where $\delta\dot{\boldsymbol{u}}$ represent the test functions and the admissible solution spaces, $\mathcal{U}^{ad}$ and $\dot{\mathcal{U}}^{ad,0}$, are defined as follows:

– $\mathcal{U}^{ad} = \left\{\boldsymbol{u} = \boldsymbol{u}(\boldsymbol{x}, t) \in [\mathcal{H}^1(\Omega_C)]^3; \boldsymbol{u} = \boldsymbol{u}_g \text{ on } \partial\Omega_C^u; \forall\, t \in [0, t_{end}]\right\}$

- $\dot{\mathcal{U}}^{ad,0} = \{\dot{\boldsymbol{u}} = \dot{\boldsymbol{u}}(\boldsymbol{x},t) \in [\mathcal{H}^1(\Omega_C)]^3; \dot{\boldsymbol{u}} = \mathbf{0} \text{ on } \partial\Omega_C^u; \forall\, t \in [0, t_{end}]\}$

3.2.1.3. *Coupling formulation*

First, the gluing region $\Omega_{GL} \subseteq \Omega_B$ where the mechanical states will be controlled must be defined. In this work, this region is chosen to be the same as the bridging region Ω_B: $\Omega_{GL} = \Omega_B$. In other words, the discrepancy between the DEM and CNEM mechanical states will be controlled in the whole Ω_B. Hereafter, the term "bridging region" will be used interchangeably to refer to Ω_{GL} or Ω_B. As mentioned before, velocity constraints are chosen to be applied in a weak sense in Ω_B. As will be seen in section 3.2.4, the velocity coupling (in which velocity constraints are applied in Ω_B) is easier and less expensive in terms of central processing unit (CPU) than the displacement coupling (in which displacement constraints are applied in Ω_B). The mediator space $\mathcal{M}$ is, therefore, defined as the space of the velocities in Ω_B. To project the discrete and continuum velocities in Ω_B onto the mediator space $\mathcal{M}$, the constrained natural neighbor (CNN) interpolation is used as projection operator Π. Since Ω_D is naturally discrete, interpolation of discrete element velocities makes no physical meaning. For computational requirements (to be able to apply the velocity constraints), the part of the discrete domain restricted to Ω_B ($\Omega_{D|\Omega_B}$) will be considered as a continuum. After defining the projection operator, it is necessary to define the junction model that must be used to enforce the DEM and CNEM compatibility in the bridging region. In this work, the $\mathcal{H}^1(\Omega_B)$ inner (scalar) product is used to apply the coupling forces ([3.5], with $(\eta_1, \eta_2) = (1, 1)$). This model which is one of the most general junction models is defined in the case of velocity coupling as:

$$< \boldsymbol{\lambda}, \dot{\boldsymbol{u}} - \dot{\boldsymbol{d}} >_{\mathcal{H}^1(\Omega_B)} = \int_{\Omega_B} \boldsymbol{\lambda} \cdot (\Pi\dot{\boldsymbol{u}} - \Pi\dot{\boldsymbol{d}}) + l^2 \boldsymbol{\varepsilon}(\boldsymbol{\lambda}) : \boldsymbol{\varepsilon}(\Pi\dot{\boldsymbol{u}} - \Pi\dot{\boldsymbol{d}})\, d\Omega \quad [3.5]$$

where $(\Pi\dot{\boldsymbol{u}} - \Pi\dot{\boldsymbol{d}})$ is the difference between the projected continuum and discrete velocities onto the mediator space $\mathcal{M}$, $\boldsymbol{\lambda}$ is the Lagrange multiplier field and l is a parameter of the $\mathcal{H}^1(\Omega_B)$ junction model, named "*junction parameter*" in this book. This parameter which has the dimension of a length is added to ensure the homogeneity of the integral terms in [3.5]. In this work, l is considered as a variable which will be studied in section 3.3. If $l = 0$, the $\mathcal{H}^1(\Omega_B)$ scalar product is reduced to the $\mathcal{L}^2(\Omega_B)$ scalar product which is well known as the Lagrange multiplier model. To partition the energies between the discrete and continuum models, three weight functions α, β and γ are introduced for, respectively, the internal energy, kinetic energy and external work in the continuum subdomain Ω_C. All these functions verify the partition of unity condition given by [1.42]. In a complementary manner, the internal energy, kinetic energy and external work in the discrete subdomain Ω_D are weighted by, respectively, $\bar{\alpha} = 1 - \alpha$, $\bar{\beta} = 1 - \beta$ and $\bar{\gamma} = 1 - \gamma$.

3.2.1.4. *Global weak formulation*

Based on the previous weak formulations of the isolated discrete and continuum subdomains, the coupling conditions can be introduced to obtain the global weak formulation of the whole problem. By introducing the weight functions in [3.2] and [3.4] and using the junction model [3.5], the global weighted weak formulation can be derived: given the initial conditions, find $(\boldsymbol{u}, \boldsymbol{d}, \boldsymbol{\theta}, \boldsymbol{\lambda}) \in \mathcal{U}^{ad} \times \mathcal{D}^{ad} \times \mathcal{O}^{ad} \times \mathcal{M}$ such that $\forall\,(\delta\dot{\boldsymbol{u}}, \delta\dot{\boldsymbol{d}}, \delta\dot{\boldsymbol{\theta}}, \delta\boldsymbol{\lambda}) \in \dot{\mathcal{U}}^{ad,0} \times \dot{\mathcal{D}}^{ad,0} \times \dot{\mathcal{O}}^{ad,0} \times \mathcal{M}$:

$$\begin{aligned}&\int_{\Omega_C} \beta\, \rho\, \delta\dot{\boldsymbol{u}} \cdot \ddot{\mathbf{u}}\, d\Omega + \int_{\Omega_C} \alpha\, \varepsilon(\delta\dot{\boldsymbol{u}}) : \boldsymbol{A} : \varepsilon(\boldsymbol{u})\, d\Omega - \int_{\partial\Omega_C^T} \gamma\, \delta\dot{\boldsymbol{u}} \cdot \boldsymbol{t}_g\, d\Gamma \\ &- \int_{\Omega_C} \gamma\, \rho\, \delta\dot{\boldsymbol{u}} \cdot \boldsymbol{f}\, d\Omega + \sum_{i=1}^{n_D} \bar{\beta}_i\, m_i\, \ddot{\boldsymbol{d}}_i \cdot \delta\dot{\boldsymbol{d}}_i + \sum_{i=1}^{n_D} \bar{\beta}_i\, I_i\, \ddot{\boldsymbol{\theta}}_i \cdot \delta\dot{\boldsymbol{\theta}}_i \\ &- \sum_{i=1}^{n_D} (\bar{\gamma}_i\, \boldsymbol{f}_i^{ext} + \bar{\alpha}_i\, \boldsymbol{f}_i^{int}) \cdot \delta\dot{\boldsymbol{d}}_i - \sum_{i=1}^{n_D} (\bar{\gamma}_i\, \boldsymbol{c}_i^{ext} + \bar{\alpha}_i\, \boldsymbol{c}_i^{int}) \cdot \delta\dot{\boldsymbol{\theta}}_i \\ &+ \delta \int_{\Omega_B} \boldsymbol{\lambda} \cdot (\Pi\dot{\boldsymbol{u}} - \Pi\dot{\boldsymbol{d}}) \,+\, l^2 \varepsilon(\boldsymbol{\lambda}) : \varepsilon(\Pi\dot{\boldsymbol{u}} - \Pi\dot{\boldsymbol{d}})\, d\Omega = 0\end{aligned} \qquad [3.6]$$

Since the last formulation [3.6] is true for any infinitesimal variation of $\dot{\boldsymbol{u}}$, $\dot{\boldsymbol{d}}$, $\dot{\boldsymbol{\theta}}$ and $\boldsymbol{\lambda}$, it can be reformulated as: given the initial conditions, find $(\boldsymbol{u}, \boldsymbol{d}, \boldsymbol{\theta}, \boldsymbol{\lambda}) \in \mathcal{U}^{ad} \times \mathcal{D}^{ad} \times \mathcal{O}^{ad} \times \mathcal{M}$ such that $\forall\,(\delta\dot{\boldsymbol{u}}, \delta\dot{\boldsymbol{d}}, \delta\dot{\boldsymbol{\theta}}, \delta\boldsymbol{\lambda}) \in \dot{\mathcal{U}}^{ad,0} \times \dot{\mathcal{D}}^{ad,0} \times \dot{\mathcal{O}}^{ad,0} \times \mathcal{M}$:

– DEM side:

$$\begin{aligned}&\sum_{i=1}^{n_D} \bar{\beta}_i\, m_i\, \ddot{\boldsymbol{d}}_i \cdot \delta\dot{\boldsymbol{d}}_i \;- \sum_{i=1}^{n_D} (\bar{\gamma}_i\, \boldsymbol{f}_i^{ext} + \bar{\alpha}_i\, \boldsymbol{f}_i^{int}) \cdot \delta\dot{\boldsymbol{d}}_i \\ &- \int_{\Omega_B} \boldsymbol{\lambda}.\delta\Pi\dot{\boldsymbol{d}} + l^2 \varepsilon(\boldsymbol{\lambda}) : \varepsilon(\delta\Pi\dot{\boldsymbol{d}})\, d\Omega = 0 \\ &\sum_{i=1}^{n_D} \bar{\beta}_i\, I_i\, \ddot{\boldsymbol{\theta}}_i \cdot \delta\dot{\boldsymbol{\theta}}_i - \sum_{i=1}^{n_D} (\bar{\gamma}_i\, \boldsymbol{c}_i^{ext} + \bar{\alpha}_i\, \boldsymbol{c}_i^{int}) \cdot \delta\dot{\boldsymbol{\theta}}_i = 0\end{aligned} \qquad [3.7]$$

– CNEM side:

$$\begin{aligned}&\int_{\Omega_C} \beta\, \rho\, \delta\dot{\boldsymbol{u}} \cdot \ddot{\mathbf{u}}\, d\Omega - \int_{\partial\Omega_C^T} \gamma\, \delta\dot{\boldsymbol{u}} \cdot \boldsymbol{t}_g\, d\Gamma + \int_{\Omega_C} \alpha\, \varepsilon(\delta\dot{\boldsymbol{u}}) : \boldsymbol{A} : \varepsilon(\boldsymbol{u})\, d\Omega \\ &- \int_{\Omega_C} \gamma\, \rho\, \delta\dot{\boldsymbol{u}} \cdot \boldsymbol{f}\, d\Omega + \int_{\Omega_B} \boldsymbol{\lambda} \cdot \delta\Pi\dot{\boldsymbol{u}} + l^2 \varepsilon(\boldsymbol{\lambda}) : \varepsilon(\delta\Pi\dot{\boldsymbol{u}})\, d\Omega = 0\end{aligned} \qquad [3.8]$$

– Interface:

$$\int_{\Omega_B} \delta\boldsymbol{\lambda} \cdot (\Pi\dot{\boldsymbol{u}} - \Pi\dot{\boldsymbol{d}}) + l^2\boldsymbol{\varepsilon}(\delta\boldsymbol{\lambda}) : \boldsymbol{\varepsilon}(\Pi\dot{\boldsymbol{u}} - \Pi\dot{\boldsymbol{d}})\, d\Omega = 0 \qquad [3.9]$$

3.2.2. *Discretization and spatial integration*

Using CNEM, the continuum subdomain Ω_C is discretized into a set of scattered nodes (with no connectivity between them). Also, since a continuous velocity field is required in $\Omega_{D|\Omega_B}$ (to apply the coupling conditions), this portion of Ω_D is considered as a continuum with regard to the velocity interpolation, and the discrete elements in this region are also assumed to act as nodes of the CNEM discretization of this continuum. It should be noted that the assumption does not affect the DEM accuracy, since the bridging region is generally located far from the fine-scale effects. In addition to Ω_C and Ω_D, another subdomain is considered in this work: the dual geometrical space of the mediator space $\mathcal{M}$. This subdomain is specific to the bridging region and will be designated hereafter by Ω_B (the same notation as this region). Contrary to the previous works on the discrete-continuum coupling, this subdomain is considered to be independent and coincides neither with $\Omega_{D|\Omega_B}$ nor with $\Omega_{C|\Omega_B}$. Consequently, the discretization of this subdomain can be made independently of the continuum and the discrete ones. This makes the application of the DEM-CNEM coupling method for complex problems easier, and provides more flexibility to the discretization of the discrete and continuum subdomains section 3.3. The new subdomain Ω_B is discretized using CNEM into a set of scattered nodes. The discretized subdomains of Ω_D, Ω_C and Ω_B are denoted by $\tilde{\Omega}_D$, $\tilde{\Omega}_C$ and $\tilde{\Omega}_B$, respectively. Since Ω_D is naturally discretized, $\tilde{\Omega}_D$ is the same as Ω_D ($\tilde{\Omega}_D = \Omega_D$). According to the configuration of the discretized subdomains in the bridging region, four cases can be distinguished (Figure 3.2). The first three configurations (Figures 3.2(a), 3.2(b) and 3.2(c)) present some regularity in the bridging region and are widely studied in the literature. In this work, the fourth configuration which is the most general one and includes the three others is retained to perform the DEM-CNEM coupling. In this configuration, no coincidence conditions are imposed in the bridging region. Therefore, it is more adapted for complex geometries. In this case, it is sufficient to discretize the different subdomains independently and mount them as indicated in Figure 3.2(d). In fact, using this configuration, it is very difficult to prove mathematically the existence and uniqueness of the solutions. Also, and contrary to the other three configurations, no numerical works studying this configuration exist in the literature. In this work, the well-posedness of the global problem will be analyzed numerically in section 3.3.

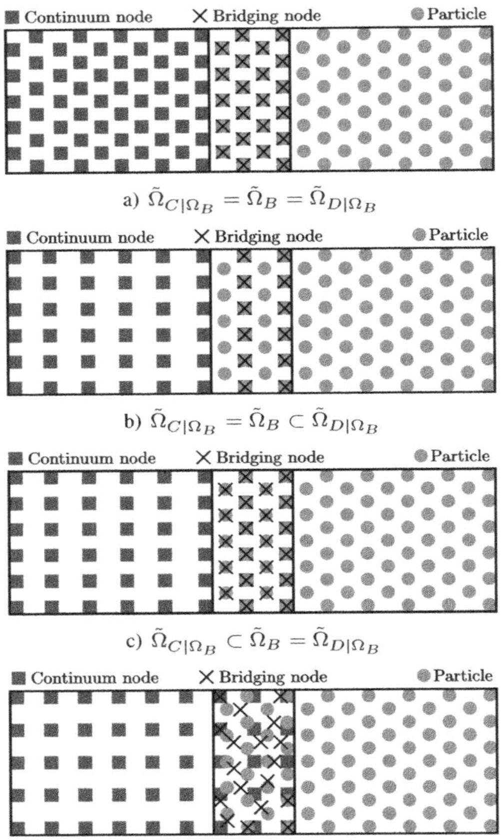

a) $\tilde{\Omega}_{C|\Omega_B} = \tilde{\Omega}_B = \tilde{\Omega}_{D|\Omega_B}$

b) $\tilde{\Omega}_{C|\Omega_B} = \tilde{\Omega}_B \subset \tilde{\Omega}_{D|\Omega_B}$

c) $\tilde{\Omega}_{C|\Omega_B} \subset \tilde{\Omega}_B = \tilde{\Omega}_{D|\Omega_B}$

d) $\tilde{\Omega}_{C|\Omega_B} \neq \tilde{\Omega}_B \neq \tilde{\Omega}_{D|\Omega_B}$

Figure 3.2. *Different configurations of the discretized subdomains in the bridging zone*

After the geometrical discretization, the discretized forms of the associated (dual) vector spaces $\mathcal{D}^{ad}$, $\mathcal{U}^{ad}$ and $\mathcal{M}$ can be derived and will be denoted hereafter by $\tilde{\mathcal{D}}^{ad}$, $\tilde{\mathcal{U}}^{ad}$ and $\tilde{\mathcal{M}}$, respectively. Using the CNN interpolation on the different discretized vector spaces, $\tilde{\mathcal{D}}^{ad}$, $\tilde{\mathcal{U}}^{ad}$ and $\tilde{\mathcal{M}}$, the displacement fields $\boldsymbol{d}$ and $\boldsymbol{u}$ and the Lagrange multipliers $\boldsymbol{\lambda}$ are approximated by:

$$\tilde{\boldsymbol{d}}(\boldsymbol{x}) = \sum_{i=1}^{n_{DB}} \phi_D^i(\boldsymbol{x})\, \boldsymbol{d}_i \qquad [3.10]$$

$$\tilde{\boldsymbol{u}}(\boldsymbol{x}) = \sum_{i=1}^{n_C} \phi_C^i(\boldsymbol{x})\, \boldsymbol{u}_i \qquad [3.11]$$

$$\tilde{\boldsymbol{\lambda}}(\boldsymbol{x}) = \sum_{i=1}^{n_B} \phi_B^i(\boldsymbol{x})\, \boldsymbol{\lambda}_i \qquad [3.12]$$

where n_{DB}, n_C and n_B are, respectively, the total number of nodes in $\tilde{\Omega}_{D|\Omega_B}$, $\tilde{\Omega}_C$ and $\tilde{\Omega}_B$. ϕ_D^i, ϕ_C^i and ϕ_B^i are the CNN shape functions defined on $\tilde{\mathcal{D}}^{ad}$, $\tilde{\mathcal{U}}^{ad}$ and $\tilde{\mathcal{M}}$, respectively. $\boldsymbol{d}_i$ are the DEM particle displacements, $\boldsymbol{u}_i$ are the CNEM nodal displacements and $\boldsymbol{\lambda}_i$ are the nodal Lagrange multipliers. Using [3.10], [3.11] and [3.12], the discretized forms of the different weak formulations [3.7], [3.8] and [3.9] can be derived. These discretized equations can be written in the form of matrix systems whose terms contain local integrals. The integrals coming from pure CNEM model and DEM model restricted to the bridging zone are evaluated numerically by the stabilized conforming nodal integration (SCNI) technique [CHE 01]. In this technique, the Voronoï cells are used as background of integration. Concerning the integrals resulting from the application of the coupling conditions, this technique cannot directly be applied. Indeed, these integrals include variables of different natures and are defined on different Voronoï diagrams. The issue here is how to choose the background of the integration in the bridging region. In this work, the Voronoï cells associated with the overlapping subdomain $\tilde{\Omega}_B$ are chosen as an integration background in this zone. All of the variables that are not defined in $\tilde{\mathcal{M}}$ are projected on this space using the CNN interpolation. After integration, the discretized equations can be derived:

– DEM side

$$\begin{aligned} \boldsymbol{m}_\beta\, \ddot{\boldsymbol{d}} &= \boldsymbol{f}_\alpha^{int} + \boldsymbol{f}_\gamma^{ext} + \boldsymbol{f}^\lambda \\ \boldsymbol{I}_\beta\, \ddot{\boldsymbol{\theta}} &= \boldsymbol{c}_\alpha^{int} + \boldsymbol{c}_\gamma^{ext} \end{aligned} \qquad [3.13]$$

with:

$\boldsymbol{m}_{\beta_{ij}} = \bar{\beta}(\boldsymbol{x}_i)\, m_i\, \delta_{ij}$: weighted mass of the discrete element (particle) i of coordinates $\boldsymbol{x}_i$.

$\boldsymbol{f}_{\alpha_i}^{int} = \bar{\alpha}(\boldsymbol{x}_i)\, \boldsymbol{f}_i^{int}$: weighted internal forces applied on particle i by its neighbors.

$\boldsymbol{f}_{\gamma_i}^{ext} = \bar{\gamma}(\boldsymbol{x}_i)\, \boldsymbol{f}_i^{ext}$: weighted external forces applied on particle i.

$\boldsymbol{I}_{\beta_{ij}} = \bar{\beta}(\boldsymbol{x}_i)\, \delta_{ij}\, I_i$: weighted mass moment of inertia of particle i.

$\boldsymbol{c}_{\alpha_i}^{int} = \bar{\alpha}(\boldsymbol{x}_i)\,\boldsymbol{c}_i^{int}$: weighted internal torques applied on particle i by its neighbors.

$\boldsymbol{c}_{\gamma_i}^{ext} = \bar{\gamma}(\boldsymbol{x}_i)\,\boldsymbol{c}_i^{ext}$: weighted external torques applied on particle i

$\boldsymbol{f}^{\lambda} = \left[\boldsymbol{c}_D^{L2} + l^2\,\boldsymbol{c}_D^{H1}\right]\boldsymbol{\lambda} = \boldsymbol{c}_D\,\boldsymbol{\lambda}$: coupling forces applied on the particles belonging to the bridging region. $\boldsymbol{c}_D^{L2}$ and $\boldsymbol{c}_D^{H1}$ are the discrete $L2$ and $H1$ coupling matrices.

– CNEM side

$$\boldsymbol{M}_{\beta}\,\ddot{\boldsymbol{u}} = -\boldsymbol{F}_{\alpha}^{int} + \boldsymbol{F}_{\gamma}^{ext} - \boldsymbol{F}^{\lambda} \qquad [3.14]$$

with:

$M_{\beta_{ij}} = \beta(\boldsymbol{x}_i)\,M_i\,\delta_{ij}$: weighted lumped mass of the node i of coordinates $\boldsymbol{x}_i$.

$\boldsymbol{F}_{\alpha}^{int} = \boldsymbol{K}_{\alpha}\,\boldsymbol{u}$: vector of weighted internal forces, where $\boldsymbol{K}_{\alpha}$ is the weighted stiffness matrix.

$\boldsymbol{F}_{\gamma}^{ext}$: vector of weighted external forces.

$\boldsymbol{F}^{\lambda} = \left[\boldsymbol{C}_C^{L2} + l^2\,\boldsymbol{C}_C^{H1}\right]\boldsymbol{\lambda} = \boldsymbol{C}_C\,\boldsymbol{\lambda}$ is the vector of coupling forces. $\boldsymbol{C}_C^{L2}$ and $\boldsymbol{C}_C^{H1}$ are the continuum $L2$ and $H1$ coupling matrices.

– Interface

$$\boldsymbol{C}_B\,\dot{\boldsymbol{u}} - \boldsymbol{c}_B\,\dot{\boldsymbol{d}} = \boldsymbol{0} \qquad [3.15]$$

with: $\boldsymbol{C}_B = {}^t\left[\boldsymbol{C}_C^{L2}\right] + l^2\,{}^t\left[\boldsymbol{C}_C^{H1}\right]$ and $\boldsymbol{c}_B = {}^t\left[\boldsymbol{c}_D^{L2}\right] + l^2\,{}^t\left[\boldsymbol{c}_D^{H1}\right]$

3.2.3. *Time integration*

The matrix systems [3.13], [3.14] and [3.15] are in the form of time-dependent ordinary differential equations (ODEs). These equations can be solved numerically using a time integration scheme. Several types of time integration schemes can be found in the literature. They can be classified into three headings: explicit, implicit and semi-implicit (or also semi-explicit). The explicit schemes calculate the state of a system at the current time step from its state at the previous time step. The implicit schemes use both the previous and current states to find the solution at the current time step. The semi-implicit schemes combine some features of the explicit and implicit schemes to find the solution. In dynamic studies, the explicit schemes are generally more recommended for capturing the fine-scale effects and for ensuring a proper momentum transfer. Numerous explicit schemes have been developed in the literature, such as the Runge–Kutta, position and velocity Verlet schemes. A comparison between these schemes can be found in [ROU 04]. According to this reference, the velocity Verlet scheme provides good results, and is easy to implement. Therefore, it is retained in this work to solve the above matrix systems. This scheme gives an $O(\Delta t^3)$ approximation for both velocities and displacements. Thus, the choice of a velocity coupling (control of velocities in the bridging region) does not affect the coupling approach accuracy compared to the displacement coupling (control of displacements in the bridging region).

3.2.4. *Algorithmic*

Now, the different ingredients for a numerical computation are well-designed and defined. Therefore, the associated algorithms that will be translated into computer codes in C++ and python languages can be derived. The key idea of these algorithms is: at each time step, ignoring the coupling conditions, predictive accelerations and velocities are determined for the two models DEM and CNEM; these quantities are then sent to a DEM-CNEM Interface (DC-Interface) code to be corrected taking into account the coupling forces; finally, the corrected quantities will be returned to the discrete and continuum codes to continue computation. This process is repeated until the simulation is achieved.

3.2.4.1. *DEM resolution*

– Initialization of $\ddot{\boldsymbol{d}}_n$, $\dot{\boldsymbol{d}}_n$, $\boldsymbol{d}_n$, $\ddot{\boldsymbol{\theta}}_n$, $\dot{\boldsymbol{\theta}}_n$ and $\boldsymbol{\theta}_n$: initial conditions or DC-Interface results.

– Computation of $\boldsymbol{d}_{n+1}$ and $\boldsymbol{\theta}_{n+1}$ using the velocity Verlet scheme:

$$\begin{aligned} \boldsymbol{d}_{n+1} &= \boldsymbol{d}_n + \Delta t\, \dot{\boldsymbol{d}}_n + \tfrac{\Delta t^2}{2}\, \ddot{\boldsymbol{d}}_n \\ \boldsymbol{\theta}_{n+1} &= \boldsymbol{\theta}_n + \Delta t\, \dot{\boldsymbol{\theta}}_n + \tfrac{\Delta t^2}{2}\, \ddot{\boldsymbol{\theta}}_n \end{aligned} \qquad [3.16]$$

– Computation of $\boldsymbol{f}^{int}_{\alpha\, n+1}$, $\boldsymbol{f}^{ext}_{\gamma\, n+1}$, $\boldsymbol{c}^{int}_{\alpha\, n+1}$ and $\boldsymbol{c}^{ext}_{\gamma\, n+1}$

– Computation of the predictive linear accelerations $\ddot{\boldsymbol{d}}^{*}_{n+1}$ (omitting the coupling forces $\boldsymbol{f}^{\lambda}_{n+1}$ from equation [3.13]).

$$\ddot{\boldsymbol{d}}^{*}_{n+1} = [\boldsymbol{m}_\beta]^{-1} \left(\boldsymbol{f}^{int}_{\alpha\, n+1} + \boldsymbol{f}^{ext}_{\gamma\, n+1}\right) \qquad [3.17]$$

– Computation of the angular accelerations $\ddot{\boldsymbol{\theta}}_{n+1}$:

$$\ddot{\boldsymbol{\theta}}_{n+1} = [\boldsymbol{I}_\beta]^{-1} \left(\boldsymbol{c}^{int}_{\alpha\, n+1} + \boldsymbol{c}^{ext}_{\gamma\, n+1}\right) \qquad [3.18]$$

– Computation of the predictive linear velocities $\dot{\boldsymbol{d}}^*_{n+1}$:

$$\dot{\boldsymbol{d}}^*_{n+1} = \dot{\boldsymbol{d}}_n + \frac{\Delta t}{2}\left(\ddot{\boldsymbol{d}}_n + \ddot{\boldsymbol{d}}^*_{n+1}\right) \quad [3.19]$$

– Computation of the angular velocities $\dot{\boldsymbol{\theta}}_{n+1}$:

$$\dot{\boldsymbol{\theta}}_{n+1} = \dot{\boldsymbol{\theta}}_n + \frac{\Delta t}{2}\left(\ddot{\boldsymbol{\theta}}_n + \ddot{\boldsymbol{\theta}}_{n+1}\right) \quad [3.20]$$

– Transfer of the predictive linear velocities and accelerations to the DC-Interface: $\dot{\boldsymbol{d}}^*_{n+1}$ and $\ddot{\boldsymbol{d}}^*_{n+1}$

3.2.4.2. *CNEM resolution*

– Initialization $\ddot{\boldsymbol{u}}_n$, $\dot{\boldsymbol{u}}_n$ and $\boldsymbol{u}_n$: initial conditions or the DC-Interface results.

– Computation of $\boldsymbol{u}_{n+1}$:

$$\boldsymbol{u}_{n+1} = \boldsymbol{u}_n + \Delta t\, \dot{\boldsymbol{u}}_n + \frac{\Delta t^2}{2}\, \ddot{\boldsymbol{u}}_n \quad [3.21]$$

– Computation of the predictive linear accelerations $\ddot{\boldsymbol{u}}^*_{n+1}$: (omitting the coupling forces $\boldsymbol{F}^{\lambda}_{n+1}$ from equation [3.14]).

$$\ddot{\boldsymbol{u}}^*_{n+1} = [\boldsymbol{M}_\beta]^{-1}\left(-\boldsymbol{F}^{int}_{\alpha\, n+1} + \boldsymbol{F}^{ext}_{\gamma\, n+1}\right) \quad [3.22]$$

– Computation of the predictive linear velocities $\dot{\boldsymbol{u}}^*_{n+1}$:

$$\dot{\boldsymbol{u}}^*_{n+1} = \dot{\boldsymbol{u}}_n + \frac{\Delta t}{2}\left(\ddot{\boldsymbol{u}}_n + \ddot{\boldsymbol{u}}^*_{n+1}\right) \quad [3.23]$$

– Transfer of the predictive linear velocities and accelerations to the DC-Interface: $\dot{\boldsymbol{u}}^*_{n+1}$ and $\ddot{\boldsymbol{u}}^*_{n+1}$.

3.2.4.3. *Interface resolution*

– Recovery of the predictive linear velocities from both the CNEM and DEM codes: $\dot{\boldsymbol{u}}^*_{n+1}$ and $\dot{\boldsymbol{d}}^*_{n+1}$

– Computation of $\boldsymbol{\lambda}_{n+1}$

$$\begin{aligned} \dot{\boldsymbol{u}}_{n+1} &= \dot{\boldsymbol{u}}^*_{n+1} - \tfrac{\Delta t}{2}\, [\boldsymbol{M}_\beta]^{-1}\, \boldsymbol{F}^{\lambda}_{n+1} \\ \dot{\boldsymbol{d}}_{n+1} &= \dot{\boldsymbol{d}}^*_n + \tfrac{\Delta t}{2}\, [\boldsymbol{m}_\beta]^{-1}\, \boldsymbol{f}^{\lambda}_{n+1} \end{aligned} \quad [3.24]$$

$$\begin{aligned} \boldsymbol{F}^{\lambda}_{n+1} &= \boldsymbol{C}_C\, \boldsymbol{\lambda}_{n+1} \\ \boldsymbol{f}^{\lambda}_{n+1} &= \boldsymbol{c}_D\, \boldsymbol{\lambda}_{n+1} \end{aligned} \qquad [3.25]$$

$$\boldsymbol{C}_B\, \dot{\boldsymbol{u}}_{n+1} - \boldsymbol{c}_B\, \dot{\boldsymbol{d}}_{n+1} = 0 \qquad [3.26]$$

Introducing equations [3.24] and [3.25] into equation [3.26], the interface system of equations can be written as:

$$\boldsymbol{A}\, \boldsymbol{\lambda}_{n+1} = \boldsymbol{b}_{n+1} \qquad [3.27]$$

where the coupling matrix $\boldsymbol{A}$ and $\boldsymbol{b}_{n+1}$ are, respectively, defined as:

$$\boldsymbol{A} = \frac{\Delta t}{2} \left(\boldsymbol{C}_B \left[\boldsymbol{M}_\beta\right]^{-1} \boldsymbol{C}_C + \boldsymbol{c}_B \left[\boldsymbol{m}_{\bar\beta}\right]^{-1} \boldsymbol{c}_D \right) \qquad [3.28]$$

$$\boldsymbol{b}_{n+1} = \boldsymbol{C}_B\, \dot{\boldsymbol{u}}^*_{n+1} - \boldsymbol{c}_B\, \dot{\boldsymbol{d}}^*_{n+1} \qquad [3.29]$$

Solving equation [3.27], $\boldsymbol{\lambda}_{n+1}$ can be obtained.

– Computation of $\boldsymbol{F}^{\lambda}_{n+1}$ and $\boldsymbol{f}^{\lambda}_{n+1}$ using equation [3.25].

– Computation of the linear velocities $\dot{\boldsymbol{u}}_{n+1}$ and $\dot{\boldsymbol{d}}_{n+1}$ using equation [3.24].

– The linear acceleration corrections: $\ddot{\boldsymbol{u}}_{n+1}$ and $\ddot{\boldsymbol{d}}_{n+1}$:

$$\begin{aligned} \ddot{\boldsymbol{u}}_{n+1} &= \ddot{\boldsymbol{u}}^*_{n+1} - \left[\boldsymbol{M}_\beta\right]^{-1} \boldsymbol{F}^{\lambda}_{n+1} \\ \ddot{\boldsymbol{d}}_{n+1} &= \ddot{\boldsymbol{d}}^*_{n+1} + \left[\boldsymbol{m}_{\bar\beta}\right]^{-1} \boldsymbol{f}^{\lambda}_{n+1} \end{aligned} \qquad [3.30]$$

– Transfer of $\ddot{\boldsymbol{u}}_{n+1}$ and $\dot{\boldsymbol{u}}_{n+1}$ to the CNEM code and transfer of $\ddot{\boldsymbol{d}}_{n+1}$ and $\dot{\boldsymbol{d}}_{n+1}$ to the DEM code.

3.2.4.4. *Remarks*

– To compute the predictive accelerations, the lumped mass matrices must be invertible. Thus, the weight functions β and $\bar\beta$ must be strictly positive in Ω_O and at the border $\partial\Omega_O$. A small ε will be used instead of zero at the nodes belonging to $\partial\Omega_O$ [1.50]. The choice of the ε value will be discussed in section 3.3.

– To determine the Lagrange multipliers vector at a later time $\boldsymbol{\lambda}_{n+1}$, the resolution of the matrix system [3.27] is performed using the well-known *LU* decomposition method [MEY 88]. In this system, the coupling matrix $\boldsymbol{A}$ [3.28] can directly be obtained. The mass matrices of both discrete and continuum models are diagonal. It is, therefore, easy to obtain their inverse matrices and to compute $\boldsymbol{A}$.

– From an algorithmic point of view, the velocity coupling used in this work is easier and less costly than the displacement coupling. Indeed, the displacement coupling requires to compute predictive displacements, in addition to the predictive accelerations and velocities, and to send them to the DC-Interface code for correction.

– No coupling conditions are explicitly applied to correct the angular velocities and angular accelerations of the discrete elements in the bridging zone. These quantities are corrected implicitly through the correction of the linear velocities and accelerations. Indeed, the internal forces are computed accounting for discrete element displacements and rotations. These forces are later used to compute the new linear velocities and accelerations which are corrected using the coupling conditions [3.15].

3.2.5. *Implementation*

Two existing workbenches are retained to perform the DEM and CNEM computation, which are, respectively, GRANular Object Oriented (GranOO) and NEtwork Solver SYstem (NESSY). GranOO is developed at the *Mechanics and Engineering Institute of Bordeaux* (I2M) by J. L. Charles *et al.* [AND 14]. This code provides C++ Object Oriented (OO) libraries to describe and solve dynamic mechanical problems by DEM using explicit temporal integration schemes. This workbench enables customization by adding plug-ins. To implement the DEM-side coupling algorithm in this code, several plug-ins are written in C++ language using the advanced OO architecture. NESSY is developed at the *Process and Engineering in Mechanics and Materials* laboratory (PIMM) by P. Lorong *et al.* This workbench provides C++ libraries interfaced with Python moduli. The CNEM-side coupling algorithm is, therefore, implemented in Python language. The communication between the two codes is ensured by "DC-Interface" developed in this work. DC-Interface which is written in Python language communicates directly with NESSY using Python classes. Whereas, the communication between DC-Interface and GranOO is performed by an inter process communication (IPC) technique. The IPC techniques can be divided into mainly four classes: message passing, inter process synchronization, shared memory and remote procedure calls. In this work, the inter process synchronization is retained. Using this technique, the synchronization of the coupled codes is automatically ensured at each time step, which allows these codes to intimately interact as in a single code. Figure 3.3 gives the overall architecture of the coupling approach.

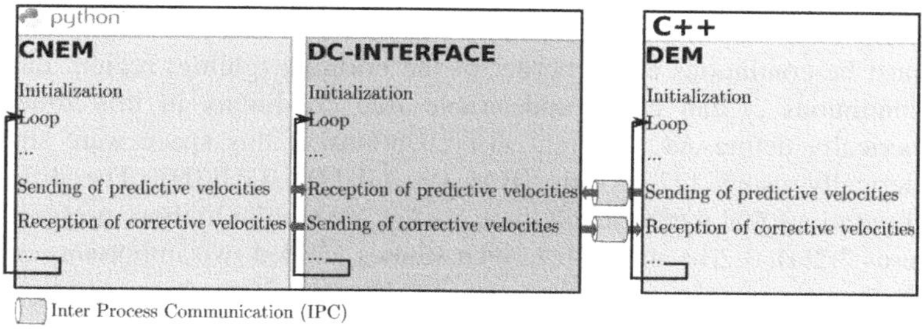

Figure 3.3. *Coupling architecture*

3.3. Parametric study of the coupling parameters

The previous section detailed how the DEM-CNEM coupling approach is performed. This approach is based on the BD technique which is well adapted for highly dynamic problems. This technique allows for avoiding spurious wave reflections at the interface between the coupled models without any additional handling, provided that its coupling parameters are well set out. Several works studying these parameters can be found in the literature. Ben Dhia and Rateau [RAT 03, BEN 01, BAU 08] were the first to study analytically these ingredients for continuum-continuum coupling. Their study is still valuable for discrete-continuum coupling. The main results concerning the well posedness of the coupling problem are recalled hereafter. The weight function α, applied to weight the internal energy, must be strictly positive in the bridging zone. Without this condition, the coercivity of the internal energy cannot be verified. Another significant result concerning the junction models is that, contrary to the $H1$ junction model ([3.5], with $(\eta_1, \eta_2) = (1, 1)$) which yields a well-posed problem, the $L2$ model ($l = 0$) can lead to an ill-conditioned system of equations, especially in the case of highly refined discretization. In the same context, Bauman *et al.* [BAU 08] have studied another junction model called $H1$ semi-norm (associated with the semi-inner product given by [3.5] with $(\eta_1, \eta_2) = (0, 1)$). This model leads to a well-posed problem, but it does not constrain enough the continuum and discrete control variables in the bridging zone. Concerning the choice of the mediator space, Ben Dhia *et al.* [BEN 08, BEN 01] mentioned that, in the case of continuous domains, it is convenient to use $\mathcal{M} = \mathcal{H}^1(\Omega_B)$; however, it is very difficult to analytically define the approximated space $\tilde{\mathcal{M}}$. To overcome the limitations of the analytical approach, other researchers [BAU 08, GUI 07, GUI 09] proposed numerical studies of the BD coupling parameters. Using static $1D$ model, Guidault *et al.* [GUI 07, GUI 09] demonstrated that, in the case of the $L2$ coupling model, the weight function α must be continuous at the border of the bridging (gluing) region. Indeed, discontinuous α can cause undesirable free conditions in this zone. To numerically define $\tilde{\mathcal{M}}$, different discretizations of this space were studied numerically using $1D$ models [BAU 08, GUI 07, GUI 09]. The different configurations that were used in these studies are similar to those presented in Figures 3.2(a), 3.2(b) and 3.2(c). Such studies yielded two important results: (1) in the case of a fine mediator space $\tilde{\mathcal{M}}$ (Figure 3.2(c)), the response of the structure does not depend on the weight functions and a locking phenomenon takes place, i.e. the fine solution exactly conforms to the coarse solution in the bridging zone; (2) in the case of a coarse mediator space $\tilde{\mathcal{M}}$ (Figure 3.2(b)), the weight functions have an influence on the solutions such that the larger the weight function on the fine-scale model, the smaller the maximum solution jump between the two models becomes.

This section aims to extend these works to $3D$ dynamic studies using the general configuration given in Figure 3.2(d). It will be demonstrated that some of the results proven in statics using $1D$ models are not valid for $3D$ dynamic simulations. Assuming the general case of the approximated mediator space (Figure 3.2(d)), the various coupling parameters studied in this work are (Figure 3.4):

– the junction parameter l [3.5];

– the weight functions α, β and γ;

– the width of the bridging region L_B;

– the discretization of the approximated mediator space $\tilde{\mathcal{M}}$ (or its associated geometric discretization characteristic length h_B).

A $3D$ beam model is used for the dynamic study (Figure 3.5), of which the length and diameter are, respectively, $L = 20\,mm$ and $D = 2\,mm$. This model is divided into two subdomains with a bridging zone. The left subdomain is modeled by the CNEM approach using 626 nodes (the associated characteristic length is about $h_C = 0.47\,mm$) and fixed at the left end ($x = 0$). The right subdomain is modeled by the DEM approach using 20 000 spherical particles having $r^{mean}_{particle} = 0.05\,mm$ as average radius. Based on the discretization characteristic lengths of DEM and CNEM subdomains ($h_D \approx r^{mean}_{particle}$ and h_C), the cutoff frequencies of the two coupled models can be determined: $f^c_C = 1.9\,MHz$ and $f^c_D = 18.2\,MHz$ ($f^c_D \approx 10\,f^c_C$). To generate a state of HFWs in propagation within the beam and investigate the wave reflexion at the DEM-CNEM interface, the free end ($x = L$) is submitted to a tensile loading with a very steep slope (Figure 3.6). As shown in Figure 3.6(b), the

Fourier spectrum contains powerful HFWs (greater than f_C^c). The material of the beam is assumed to be silica glass, whose the mechanical properties are given in Figure 3.5. The corresponding microscopic properties of the cohesive beam bonds between the discrete elements in the DEM model are given in Table 3.1. For the sake of simplicity, the same weight function is used to weight the different energies ($\alpha = \beta = \gamma$). To control the wave propagation within the model, four check points are placed along this beam (Figure 3.5) as follows:

– CnemCheckPoint: at the middle of the CNEM subdomain where the controlled quantities are computed using the CNEM nodes in this zone.

– OverlapCnemCheckPoint: at the middle of the bridging zone where the controlled quantities are computed using only the CNEM nodes in this zone.

– OverlapDemCheckPoint: at the middle of the bridging zone where the controlled quantities are computed using only the DEM particles in this zone.

– DemCheckPoint: at the middle of the DEM subdomain where the controlled quantities are computed using the DEM particles in this zone.

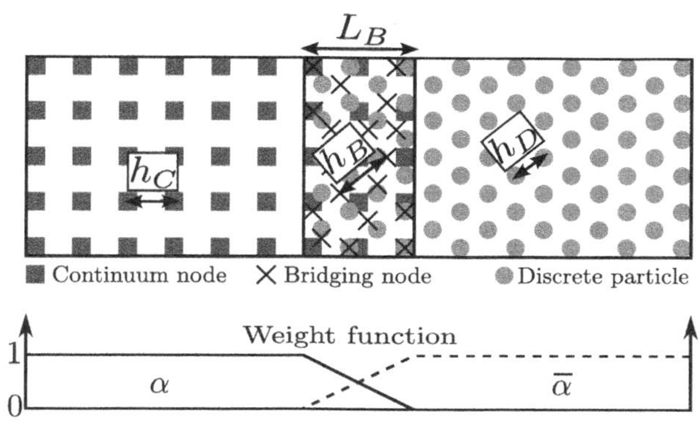

Figure 3.4. *Main parameters of the bridging domain coupling approach*

Young's modulus	Poisson's ratio	Dimensionless radius
$E_\mu = 265\,GPa$	$\nu_\mu = 0.3$	$r_\mu = \frac{r_{beam}}{r_{particle}^{mean}} = 0.71$

Table 3.1. *The microscopic properties of the cohesive beam bonds in the discrete subdomain*

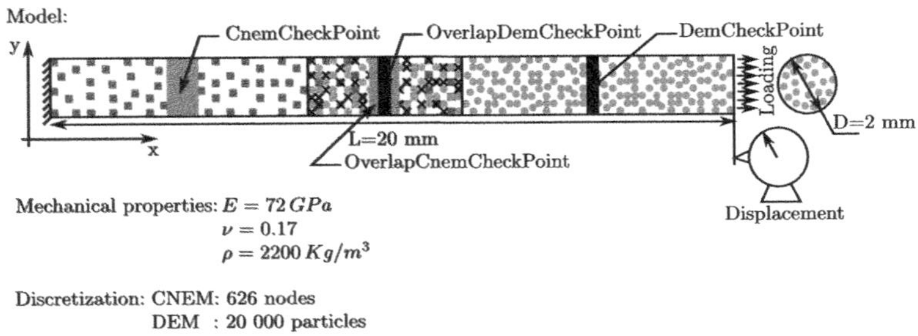

Figure 3.5. *Beam model for the parametric study*

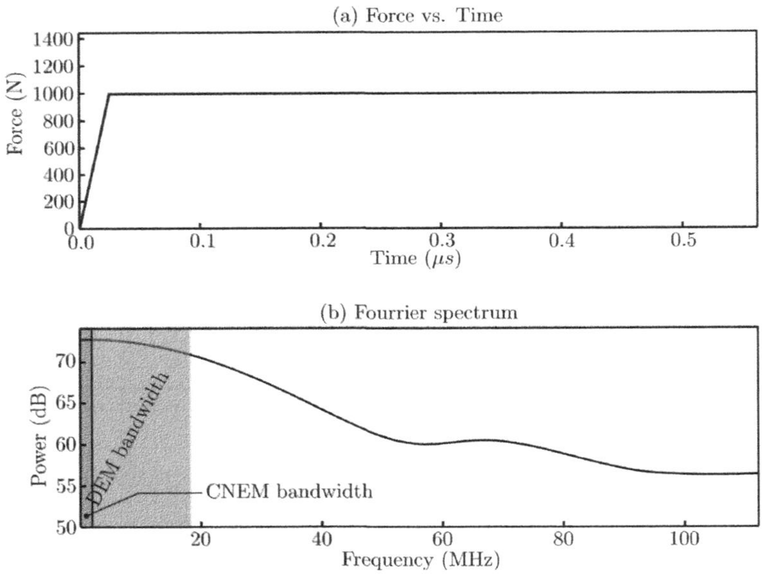

Figure 3.6. *Tensile loading used for the parametric study and its spectral analysis (computed from FFT)*

Before studying the coupling parameters, it is important to ensure that the models to be coupled are equivalent. To this end, tensile tests of the $3D$ beam presented in Figure 3.5 are first simulated using fully DEM computation and fully CNEM computation. Figure 3.7 shows the associated results. Table 3.2 presents the mean displacement of the right end and the first three natural frequencies. It can be seen that the results are in good agreement, and they are also in agreement with the beam theory results. This ensures the equivalence of the two models.

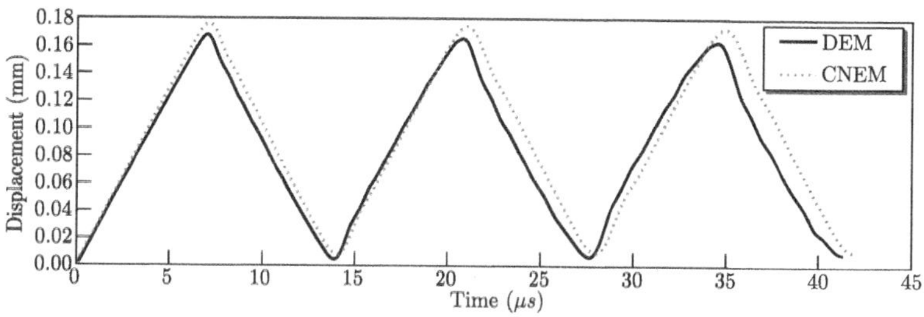

Figure 3.7. *Free-end displacement obtained by fully DEM computation and fully CNEM computation*

	U_{mean} (mm)	f_0 (Hz)	f_1 (Hz)	f_2 (Hz)
Theory	0.087	71 757	215 272	358 787
DEM	0.083	72 408	217 246	362 072
CNEM	0.088	71 359	214 023	356 491

Table 3.2. *Comparison of DEM, CNEM and analytical results*

3.3.1. *Influence of the junction parameter l*

This parameter, arising from the application of the $\mathcal{H}^1(\Omega_B)$ junction model ([3.5], with $(\eta_1, \eta_2) = (1, 1)$), is mainly employed to compute the global coupling matrix $\boldsymbol{A}$ [3.28]. The influence of this parameter l on the coupling results will, therefore, be investigated by studying its effect on the conditioning of $\boldsymbol{A}$:

$$\mathrm{Cond}(\boldsymbol{A}) = \parallel \boldsymbol{A} \parallel . \parallel \boldsymbol{A}^{-1} \parallel \qquad [3.31]$$

which is an important indicator of the results accuracy. A coarse mediator space $\tilde{\mathcal{M}}$ (coarse $\tilde{\Omega}_B$, i.e. at the same order as $\tilde{\Omega}_C$) and continuous weight functions with $\varepsilon = 0.05$ (ε is defined in [1.50]) are used in this study. Figure 3.8 shows the conditioning of $\boldsymbol{A}$ with respect to l. The $\boldsymbol{A}$ conditioning decreases with l and reaches a minimum for $l = l_{opt}$. Beyond this value, the conditioning increases exponentially as l increases. Therefore, the $H1$ coupling (based on the $\mathcal{H}^1(\Omega_B)$ junction model) for a small value of l is better than the $L2$ coupling ($l = 0$). However, contrary to what is often presented in the literature, the $H1$ coupling becomes worse if l exceeds some optimal value.

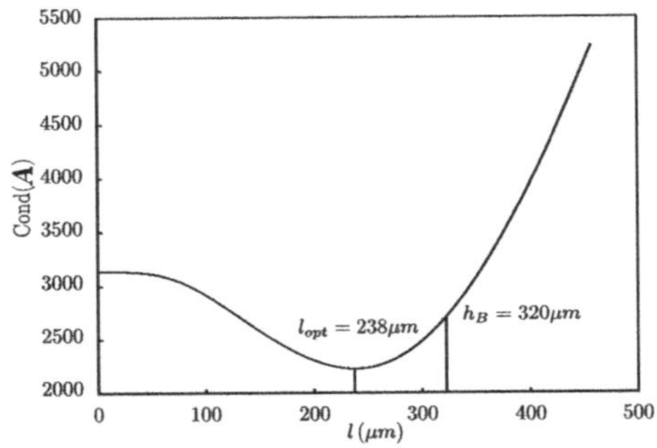

Figure 3.8. *Conditioning of $\boldsymbol{A}$ with respect to l $L_B = 6\,mm$, coarse mediator space and continuous weight functions with $\varepsilon = 0.05$*

Table 3.3 presents the optimal values of the junction parameter (l_{opt}) determined numerically for different L_B and h_B (width and discretization characteristic length of $\tilde{\Omega}_B$, respectively). l_{opt} practically does not change with the width of the bridging region, except for very fine mediator space (case of $h_B = 32\,\mu m$). However, this parameter depends significantly on h_B. In the case of relatively coarse mediator space (case of $h_B = 320\,\mu m$), l_{opt} is roughly of the same order of magnitude as h_B. In practice, the junction parameter l can be chosen as: $l = h_B$. This choice generally results in acceptable conditioning of $\boldsymbol{A}$. In the remainder of this chapter, l is set equal to l_{opt} which is determined numerically for each coupling parameters combination.

	$h_B\ (\mu m)$	$l_{opt}\ (\mu m)$
$L_B = 2\,mm$	320	230
	32	0.01
$L_B = 4\,mm$	320	230
	32	2.2
$L_B = 6\,mm$	320	238
	32	12

Table 3.3. *Variation of l_{opt} with the width L_B and discretization characteristic length h_B of the bridging subdomain*

3.3.2. *Influence of the weight function* α

The influence of the weight function α is studied in this section. The other coupling parameters are chosen as follows: a fine mediator space $\tilde{\mathcal{M}}$ (fine discretization of $\tilde{\Omega}_B$, i.e. at the same order as the discretization of $\tilde{\Omega}_D$) and $L_B = 2\,mm$.

3.3.2.1. *Constant weight functions* $\alpha = \bar{\alpha} = 0.5$

The DEM and CNEM models are supposed to be equivalent in the bridging region, i.e. weighted in the same manner (Figure 1.5(a)). Figure 3.9 presents the free-end displacement obtained in this case. The mean displacement obtained using the coupling method is $U_{mean} = 0.081\,mm$. This is in agreement with the reference results (Table 3.2). However, the temporal coupling displacement (Figure 3.9) presents several deviations with regard to the reference curves. To analyze this coupling result and to determine the cause of these deviations, the velocities at the different check points (Figure 3.5) are plotted for the first round trip of the wave propagation (Figure 3.10). This figure shows that the HFWs initially captured at the "DemCheckPoint" do not appear in "OverlapDemCheckPoint" or "OverlapCnemCheckPoint". In other words, the major part of the HFWs is reflected without entering the bridging region. This explains the deviation in the temporal displacement each time the global wave crosses the bridging region. Thus, constant weight functions are not a good choice for dynamic simulations. Indeed, the projection mechanism, which occurs in Ω_B, cannot dampen the HFWs, and additional filtering is required in this case. In contrast, in the case of static studies, Guidault *et al.* [GUI 07, GUI 09] have shown that constant weight functions can be used with $H1$ coupling.

3.3.2.2. *Constant weight functions* $\alpha \neq 0.5$

Keeping constant weight functions, the influence of the weight constant on the wave propagation is analyzed. Two cases are studied here: the first case makes the discrete model more influential in the bridging region ($\alpha = 0.3$), whereas the opposite trend is investigated in the second case ($\alpha = 0.8$) in which the continuum model is assumed to be dominating in the bridging region. The associated results are presented in Figure 3.11. A large difference between the results is observed. In the first case ($\alpha = 0.3$), the magnitude of the free-end displacement is greater than the one obtained using $\alpha = 0.5$. However, it is smaller in the case of $\alpha = 0.8$. To provide an explanation for these results, the temporal velocities at the check points are presented in Figure 3.12.

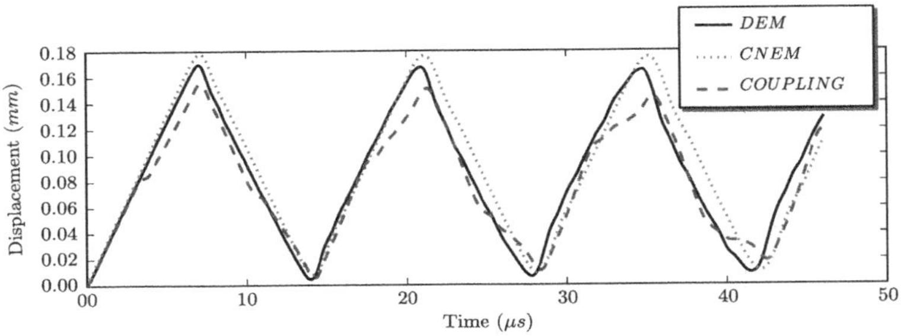

Figure 3.9. *Free-end displacement obtained using DEM, CNEM and the coupling method;* $L_B = 2\,mm$*, fine mediator space and constant weight functions* $\alpha = 0.5$

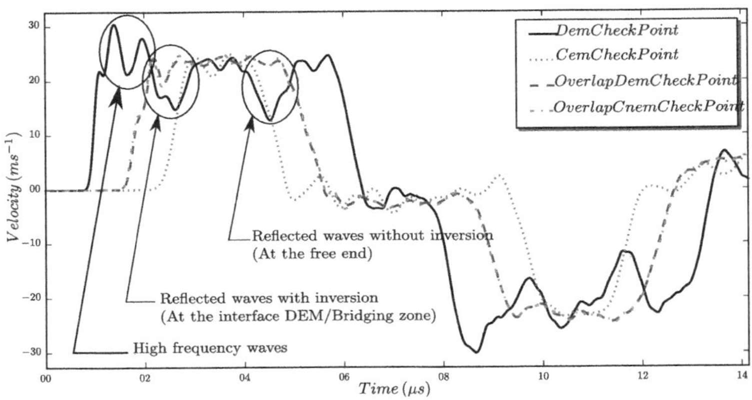

Figure 3.10. *Linear velocities at the check points for the first round trip;* $L_B = 2\,mm$*, fine mediator space and constant weight functions* $\alpha = 0.5$

For $\alpha = 0.8$, a portion of the global incident wave is positively reflected at the interface between the two models without entering the bridging zone. Furthermore, only a complementary part is transmitted in the CNEM model. Quantitatively, the transmission and reflection coefficients are $t^{num} = 0.44$ and $r^{num} = 0.56$, respectively. Analogous to the wave propagation between media with different acoustic impedances, we define the transmission and reflection coefficients in terms of α and $\bar{\alpha}$ as follows: $t^{th} = \frac{2\,\bar{\alpha}}{\alpha+\bar{\alpha}}$ and $r^{th} = \frac{\alpha-\bar{\alpha}}{\alpha+\bar{\alpha}}$. It can be verified that t^{num} and r^{num} are of the same order of magnitude as $t^{th} = 0.4$ and $r^{th} = 0.6$, respectively.

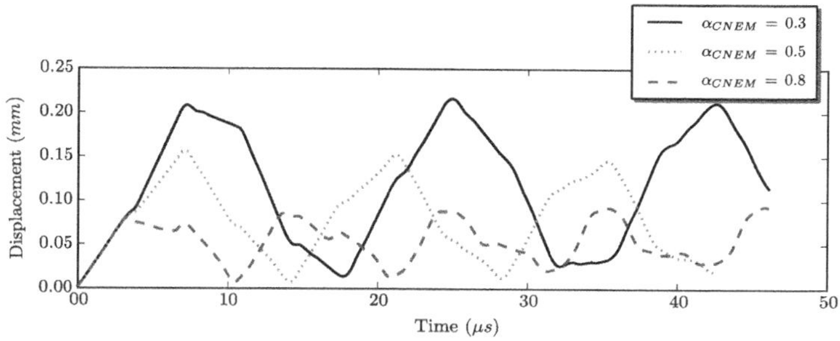

Figure 3.11. *Free-end displacement obtained using the coupling method with different constant weight functions:* $\alpha = 0.3$, $\alpha = 0.5$ *and* $\alpha = 0.8$; $L_B = 2\,mm$ *and fine mediator space*

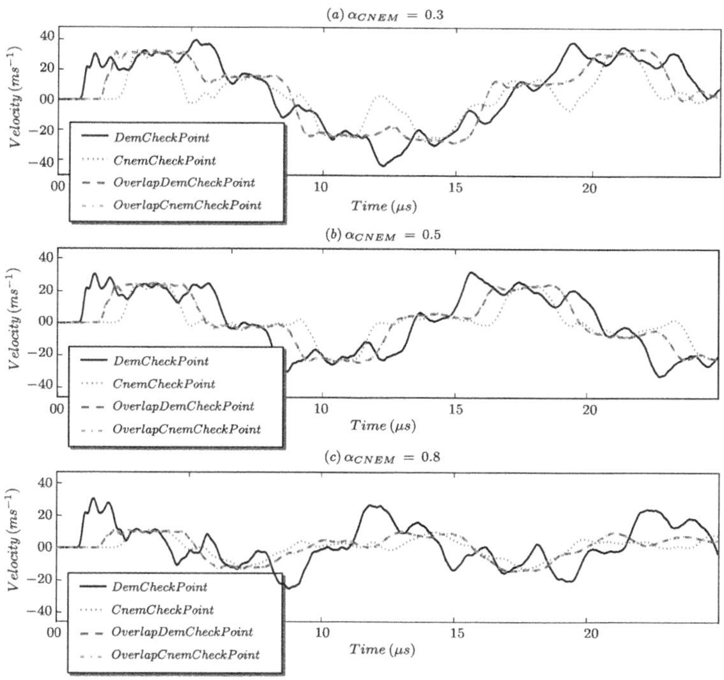

Figure 3.12. *Linear velocities at the check points for different constant weight functions:* $\alpha = 0.3$, $\alpha = 0.5$ *and* $\alpha = 0.8$; $L_B = 2\,mm$ *and fine mediator space*

For $\alpha = 0.3$, the same reflection mechanism takes place but with a negative coefficient. Indeed, the velocity magnitude of the transmitted wave (measured at "CnemCheckPoint") is greater than the velocity magnitude of the incident wave (initially measured at "DemCheckPoint"). Also, the numerical transmission and reflection coefficients, $t^{num} = 1.44$ and $r^{num} = -0.44$, are of the same order of magnitude as $t^{th} = 1.4$ and $r^{th} = -0.4$.

Therefore, for the case of constant weighting, the use of $\alpha \neq 0.5$ is worse with regard to the reflection phenomenon. In this case, a part of the global incident wave is reflected at the interface between the models (not only the HFWs which are not supported by the coarse-scale model). This result proves that the one-dimensional (1D) static studies available in the literature cannot be used to perform dynamic coupling. Guidault *et al.* [GUI 07] have shown that, in statics and using a fine mediator space, the solutions do not depend on the weight constant.

3.3.2.3. *Continuous weight functions*

As explained earlier, the weight functions must not vanish at the boundary of the bridging zone $\partial\Omega_B$ (for the weighted lumped mass matrices to be invertible), and a small value ε must be applied rather than zero at $\partial\Omega_B$. Before studying the influence of the continuous weight functions, the influence of ε is investigated. Figure 3.13 presents the free-end displacement using continuous weight functions for different ε. This parameter ε, when less than 0.05, has no practical influence on the results, but a very small ε can lead to instability problems. Indeed, as shown in Table 3.4, the smaller the ε, the greater the conditioning of the coupling matrix $\boldsymbol{A}$ becomes.

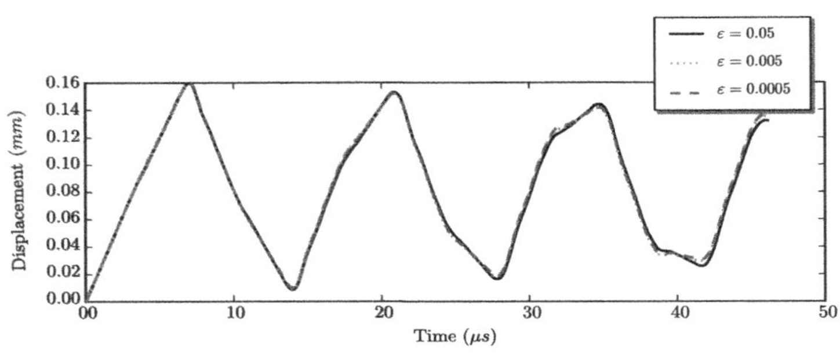

Figure 3.13. *Free-end displacement obtained using the coupling method for different values of ε; $L_B = 2\,mm$, fine mediator space and continuous weight functions*

ε	0.05	0.005	0.0005
Cond $[\boldsymbol{A}(l_{opt})]$	$2.53e4$	$8.93e4$	$5.67e5$

Table 3.4. *Conditioning of* ***A*** *with respect to* ε*;* $L_B = 2\,mm$*, fine mediator space, continuous weight functions*

In the remainder of this section, $\varepsilon = 0.05$ is used each time a continuous weight function is applied. Figure 3.14 shows the free-end displacement for the case of a linear continuous weight function (Figure 1.5(b)). The associated velocities measured at the different check points are presented in Figure 3.15. No HFWs are reflected at the interface between the two models. Using continuous weight functions, the HFWs enter the bridging zone, and then, they are dampened by the projection of the fine space solution (from the DEM model) onto the coarse space solution (from the CNEM model). Figure 3.15(b) shows that with a fine mediator space, a small bridging zone is sufficient to cancel out all of the HFWs. The use of a continuous weight function significantly improves the results (Figure 3.14). However, a small deviation from the reference results still persists and becomes greater each time the wave travels back (CNEM-DEM direction). Because of the very fine discretization of the DEM subdomain, the weight of the particles in the bridging zone decreases smoothly when approaching the CNEM domain. By examining the first round-trip in Figure 3.14, it can be observed that practically no deviation from the reference is noted when the wave travels from the DEM subdomain to the CNEM subdomain. Therefore, the incident wave correctly crosses the interface between the pure DEM ($\Omega_D \backslash \Omega_B$) and the bridging zone (Ω_B). However, in the CNEM subdomain, a coarse discretization is used. The jump between the weight values of two adjacent nodes is relatively large. Thus, the same reflection mechanism, observed using constant weight functions with $\alpha \neq 0.5$, occurs when the wave travels back (CNEM-DEM direction). To enhance the results, we can increase the width of the bridging zone to reduce the slope of the weight functions and so the weight jump between adjacent nodes. Another solution consists of using continuous differentiable weight functions (Figure 1.5(c) to reduce the weight jump in the vicinity of $\partial\Omega_B$ (boundary of the bridging zone). Figures 3.23, 3.24 and 3.25 present the results using the two solutions. The wave correctly crosses Ω_B without any significant deviation. In conclusion, the continuous differentiable weight functions (Figure 1.5(c)) are best adapted to weight the different energies in the bridging zone. These functions ensure a better regularity (smoothing) in the vicinity of the $\partial\Omega_B$, and then the HFWs cross the bridging zone where they are dampened, provided that L_B and $\tilde{\mathcal{M}}$ are well chosen. Otherwise, the same reflection mechanism can take place but for other reasons as will be explained in the next sections.

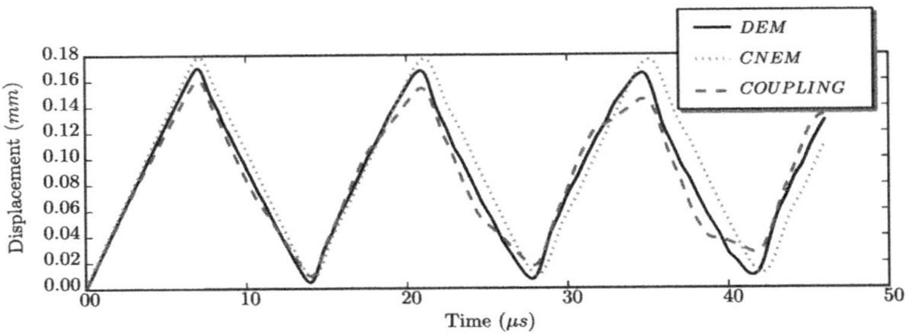

Figure 3.14. *Free-end displacement obtained using DEM, CNEM and the coupling method; $L_B = 2\,mm$, fine mediator space and continuous weight functions*

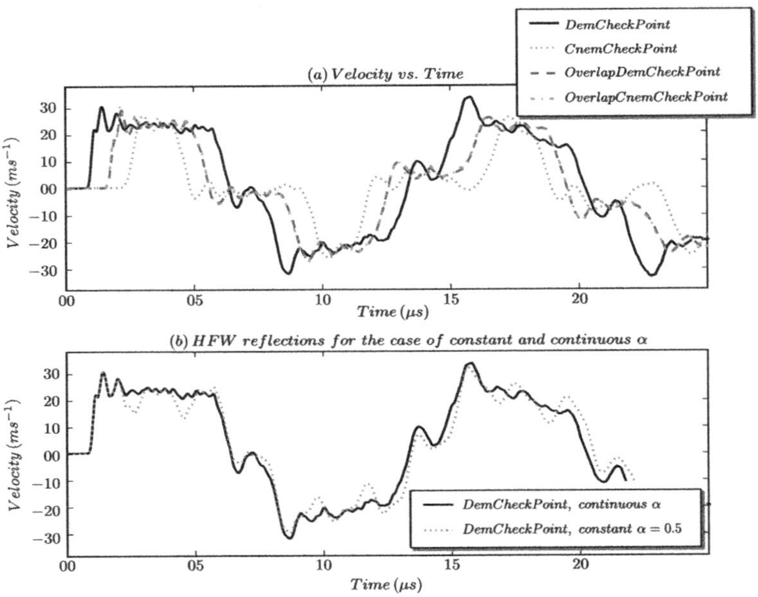

Figure 3.15. *a) Linear velocities at the different check points for the case of continuous α; b) comparison of linear velocities at the "DemCheckPoint" obtained using continuous and constant ($\alpha = 0.5$) weight functions; $L_B = 2\,mm$ and fine mediator space*

3.3.3. *Influence of the approximated mediator space* $\tilde{\mathcal{M}}$

In the previous section, a fixed fine mediator space has been used. In such a case, the velocities of fine-scale model are practically locked on those of the coarse-scale model (Figure 3.16(a)). Indeed, the velocity curve at the "OverlapDemCheckPoint" is practically superimposed on that at the "OverlapCnemCheckPoint". The same locking phenomenon is noted in the literature when the third configuration (Figure 3.2(c)) is used [GUI 07]. Now, to study the influence of $\tilde{\mathcal{M}}$ on the coupling results, a coarse mediator space (coarse discretization of $\tilde{\Omega}_B$, i.e. at the same order as the discretization of $\tilde{\Omega}_C$) is used. As shown in Figure 3.16(b), the equality of velocities in the bridging region is satisfied only in a weak sense and not at each node of the mediator space. This allows the fine model (DEM model) to correctly act in Ω_B. However, in this case, a small bridging zone is insufficient to correctly transmit the global incident wave and cancel the HFWs. A solution to reduce the wave reflections at the interface between the two models is to use a large bridging zone (large L_B). As a result of this solution, a large discrete subdomain has to be used which is costly and cumbersome. However, contrary to the previous works on coupling approaches, the discretization of the mediator space $\tilde{\mathcal{M}}$ can be made independently from the discrete and continuum models in this work. Therefore, an intermediate discretization of $\tilde{\mathcal{M}}$, between the discretization of the continuum and discrete models, can be used. As it will be seen in the next sections, for any choice of L_B, there is an optimal discretization of the mediator space h_B that cancels the HFWs in the bridging region.

3.3.4. *Influence of the width of the bridging zone* L_B

As seen in the previous section, the influence of the bridging width (L_B) on the coupling results depends on the approximated mediator space $\tilde{\mathcal{M}}$. Figures 3.17 and 3.18 provide the coupling results for different L_B using fine and coarse mediator spaces, respectively. In the case of fine $\tilde{\mathcal{M}}$, the width of the bridging region has no significant influence on the results because of the locking phenomenon. In such a case, the DEM particles are strongly constrained in the bridging region which in turn will represent a rigid wall for the HFWs, regardless of L_B. However, in the case of coarse $\tilde{\mathcal{M}}$, the discrete particles can somewhat freely act in the bridging region. Therefore, using a small L_B (narrow bridging region), the interface between the two models will act like a free surface for the HFWs. In this case, a large bridging region is required to sufficiently constrain the DEM particles in the bridging region and then to reduce the HFWs reflections. Figure 3.18 shows that, in the case of coarse mediator space, the coupling results are better as L_B increases.

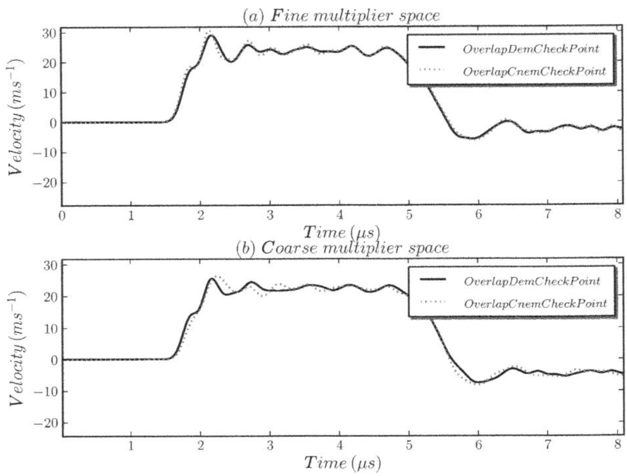

Figure 3.16. *Velocities comparison in the bridging zone using fine and coarse mediator spaces;* $L_B = 2\,mm$, *continuous weight function and* $\varepsilon = 0.05$

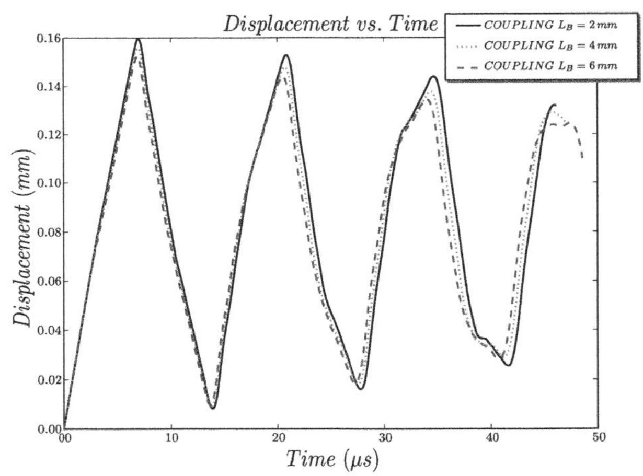

Figure 3.17. *Free-end displacements obtained using the coupling method for* $L_B = 2\,mm$, $L_B = 4\,mm$ *and* $L_B = 6\,mm$; *fine mediator space, continuous weight functions,* $\varepsilon = 0.05$

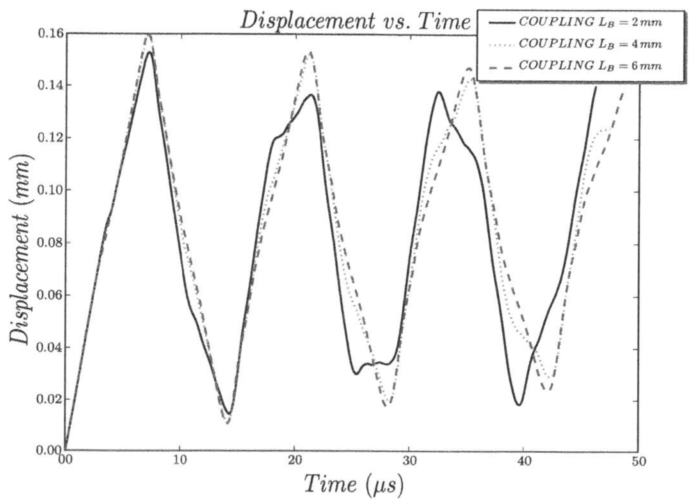

Figure 3.18. *Free-end displacements obtained using the coupling method for $L_B = 2\,mm$, $L_B = 4\,mm$ and $L_B = 6\,mm$; coarse mediator space, continuous weight functions and $\varepsilon = 0.05$*

3.3.5. *Dependence between L_B and $\tilde{\mathcal{M}}$*

This section highlights the dependence between L_B and $\tilde{\mathcal{M}}$. First, the discretization characteristic length of the bridging subdomain h_B will be investigated. The width of the bridging region is fixed at $L_B = 2\,mm$. Also, to investigate the effect of the discretization characteristic lengths of the continuum and discrete subdomains (h_C and h_D, respectively) on h_B, these parameters are first chosen as: $h_C = 5\,h_D$. Figure 3.19 shows the free-end displacements for different choices of h_B. According to this figure, the optimal value of h_B is $h_B = 2\,h_D$. Indeed, no significant deviation is pointed in the associated result. However, inward deviations and outward deviations with respect to the reference results are noted in the case of $h_B = h_D$ and $h_B = 3\,h_D$, respectively. The inward deviations report that the DEM particles are strongly constrained; whereas the outward deviations indicate that these particles are not sufficiently constrained. Using $h_C = 10\,h_D$, the same conclusions and remarks, as for $h_C = 5\,h_D$, are noted (Figure 3.20). However, in this case, the optimal value of h_B is $h_B = 3\,h_D$. This value depends, *inter alia*, on the ratio between h_D and h_C. The larger this ratio, the larger the optimal value of h_B becomes.

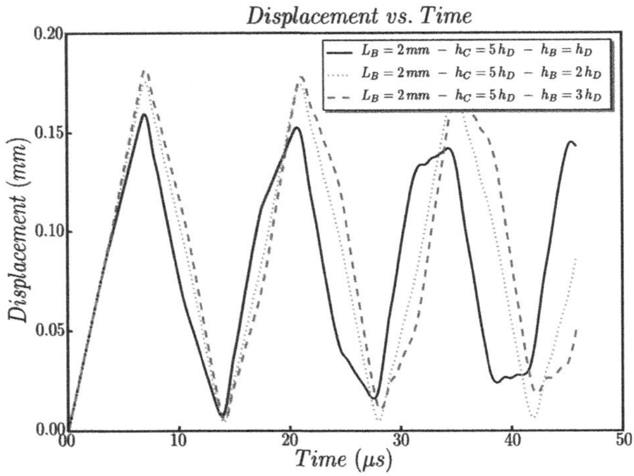

Figure 3.19. *Influence of the discretization characteristic length of the bridging subdomain h_B; $L_B = 2\,mm$ and $h_C = 5\,h_D$*

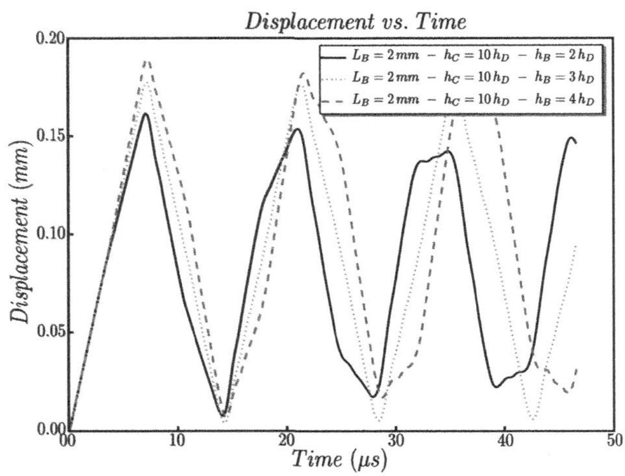

Figure 3.20. *Influence of the discretization characteristic length of the bridging subdomain h_B; $L_B = 2\,mm$ and $h_C = 10\,h_D$*

To analyze the influence of L_B on the optimal value of h_B, $L_B = 6\,mm$ is then used when $h_C = 10\,h_D$ is held fixed. Figure 3.21 presents the associated results. The same conclusions and remarks, as for $L_B = 2\,mm$, are pointed.

However, in the present case, the optimal value is $h_B = 9\,h_D$ and the deviation mechanisms are much less sensitive to h_B. In conclusion, the optimal value of h_B depends also on L_B, but the choice of this parameter is less sensitive when large bridging region is used.

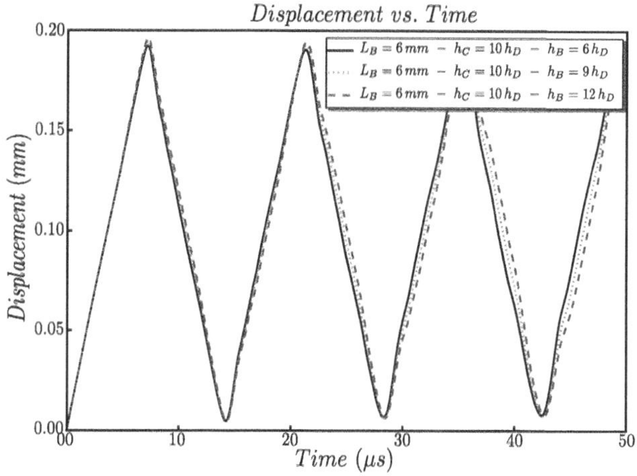

Figure 3.21. *Influence of the discretization characteristic length of the bridging subdomain h_B; $L_B = 6\,mm$ and $h_C = 10\,h_D$*

3.4. Choice of the coupling parameters in practice

In a general case, there is not an obvious method to determine, in a single way, the various coupling parameters to avoid wave reflections. This section gives several recommendations and trends to choose these parameters correctly. To minimize the conditioning of the coupling matrix $\boldsymbol{A}$, the $H1$ coupling with $l = l_{opt}$ is recommended. l_{opt} can be chosen at the some order of magnitude as the discretization characteristic length of the bridging subdomain ($l_{opt} \approx h_B$). Although, in the case of a very fine mediator space $\tilde{\mathcal{M}}$, the optimal value of l is not of the same order of magnitude as h_B, the choice of $l \approx h_B$ generally results in acceptable conditioning of $\boldsymbol{A}$. The weight functions must have at least C^0 continuity in the bridging region, but more regular functions (C^1 continuity or more) are recommended. Indeed, with constant weight functions, the HFWs are reflected without entering the bridging region and cannot be dampened by the projection mechanism which occurs in this region. The choice of L_B and $\tilde{\mathcal{M}}$ is more challenging. Indeed, these parameters depend on each other, and on the discretization of the discrete and continuum models. In practice, the choice of these parameters can be made as follows: the discrete and continuum models are discretized according to the physical effects that have to be captured in the associated regions; then, the width of the bridging region can be fixed as narrowly as possible, containing a sufficient number of CNEM nodes and DEM particles; finally, $\tilde{\mathcal{M}}$ can be chosen such that $h_D \leq h_B \leq h_C$. Using an arbitrary choice of h_B banded between h_D and h_C, a dynamic simulation can be performed to control the wave propagation between the models; if an inward deviation is found when the wave crossing the interface, this means that the coupled models are strongly constrained in the bridging region and h_B must be increased; if an outward deviation is found, this means that the coupled models are not sufficiently constrained and h_B must be decreased; otherwise, the optimal value of h_B is luckily used.

3.5. Validation

The previous parametric study highlighted several recommendations to perform a correct coupling, using $3D$ beam submitted to a dynamic tensile loading. In this section, the results of this study are used to validate the DEM-CNEM coupling approach using several elementary tests: $3D$ beams submitted to dynamic tension, dynamic bending, dynamic torsion and initial velocity loading. Contrary to the tensile loading case, in bending and torsion tests, the strains in the cross-sections are significant. To account for these effects, new geometrical characteristics of the $3D$ beam model are used: $L = 100\,mm$ and $D = 20\,mm$ ($L/D = 5$). The DEM model is only applied in a small portion located 20 mm from the right end (the section located at $x = L$) and the remainder of the domain is modeled with CNEM (Figure 3.22). The different coupling parameters are chosen as follows: $L_B = 10\,mm$, continuous differentiable weight functions, $\varepsilon = 0.05$, coarse mediator space, $l = h_B$.

Figures 3.23 and 3.24 present the temporal free-end displacements with respect to the $x - axis$ and $y - axis$ using, respectively, the tensile and bending loading. The deviation from the reference, as observed in the previous simulations when the wave crosses Ω_B, disappeared in the present results. Figure 3.25 presents the temporal free-end rotation about the $x - axis$ using the torsional loading. The coupling results are in good agreement with the DEM results (Table 3.5). Finally, the coupling method is tested using an initial velocity loading (test 4 in Figure 3.22). Figure 3.26 presents the associated free-end displacement with respect to $x - axis$. The coupling result is comparable to the reference one, obtained from fully DEM computation.

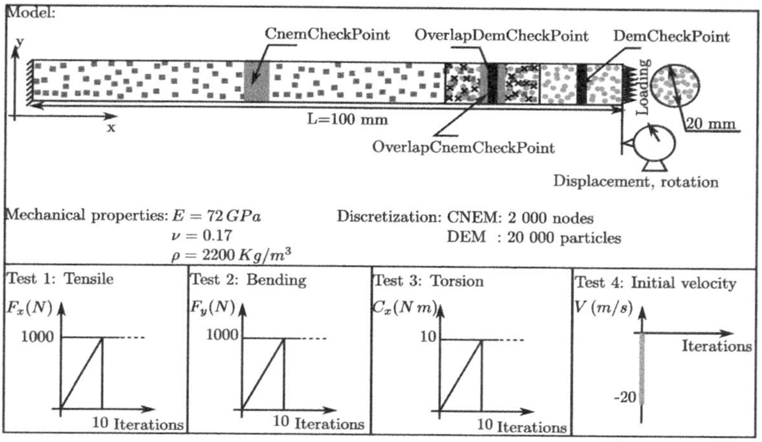

Figure 3.22. *Validation model*

		Beam theory	DEM	CNEM	Coupling
Tension	$U_{x\ mean}(mm)$	$4.40e-3$	$4.44e-3$	$4.49e-3$	$4.61e-3$
	$f_0(Hz)$	14 351	14 235	14 262	14 425
Bending	$U_{y\ mean}(mm)$	$5.85e-1$	$5.86e-1$	$6.27e-1$	$6.13e-1$
	$f_0(Hz)$	1 606	1 594	1 557	1 595
Torsion	$\theta_{x\ mean}(mm)$	$2.05e-3$	$2.34e-3$	–	$2.32e-3$
	$f_0(Hz)$	9 382	9 252	–	9 106

Table 3.5. *Comparison of the results; $L_B = 10\,mm$, coarse mediator space, continuous differentiable weight functions, $\varepsilon = 0.05$*

3.6. Conclusion

In this chapter, a three-dimensional (3D) discrete-continuum coupling method adapted for dynamic simulations was developed. This approach couples the DEM with the CNEM which was retained in Chapter 2. This coupling approach is based on the BD technique which is well suitable for problems in fast dynamics. The most general configuration of the superimposed subdomains in the bridging region (Figure 3.2(d)) was retained to perform the DEM-CNEM coupling. In this configuration, no coincidence conditions are imposed between the nodes of the different subdomains in the bridging zone. Therefore, the different subdomains can be discretized independently, which makes easier the application of this coupling approach on complex geometries. As shown, the well-posedness of the global system of equations was verified numerically.

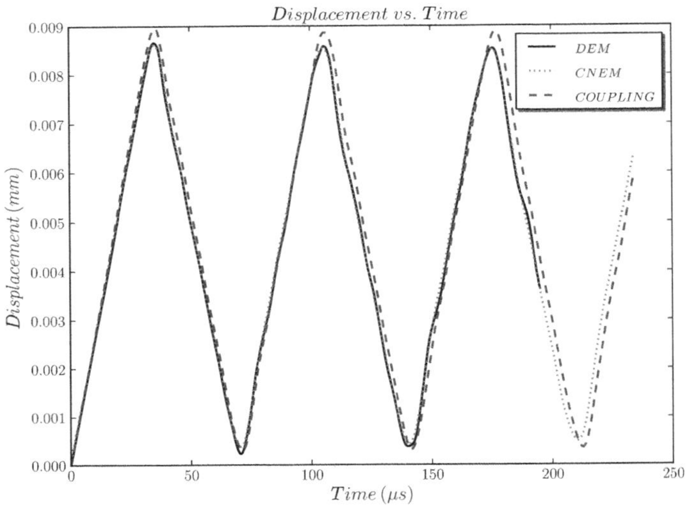

Figure 3.23. *Tensile loading: free-end displacements $U_{x\ mean}$ obtained by DEM, CNEM and the coupling method; $L_B = 10\,mm$, coarse mediator space, continuous differentiable weighting and $\varepsilon = 0.05$*

Several parameters are involved in this coupling approach: the junction parameter l, the weight functions α, β and γ, the width of the overlapping region L_B and the approximated mediator space $\tilde{\mathcal{M}}$. To simplify the setting of these parameters in practice, a parametric study of these parameters was performed using $3D$ beam model subjected to dynamic tensile loading. For the case of a small junction parameter l, the $H1$ coupling is more accurate than the $L2$ coupling. Indeed, it decreases the conditioning of the coupling matrix $\boldsymbol{A}$. However, beyond a small value (l_{opt}), it can lead to instability problems. Therefore, it is important to carefully choose this parameter. In practice, l can be chosen as the discretization characteristic length of the overlapping subdomain ($l = h_B$). This choice generally results in acceptable conditioning of $\boldsymbol{A}$. The weight functions must have at least C^0 continuity in the bridging region, but more regular functions (having C^1 continuity or more) are recommended. Indeed, with constant weight functions, the HFWs are reflected without entering the overlapping region and cannot be dampened by the projection mechanism which occurs in this region. The choice of L_B

and $\tilde{\mathcal{M}}$ is more challenging. These two parameters depend on each other, and on the discretization of the discrete and continuum subdomains. In practice, L_B can first be fixed such that the bridging region contains sufficient number of DEM elements and CNEM nodes, and then preliminary dynamic simulations using simple dynamic loading and different $\tilde{\mathcal{M}}$ can be performed to choose the appropriate one.

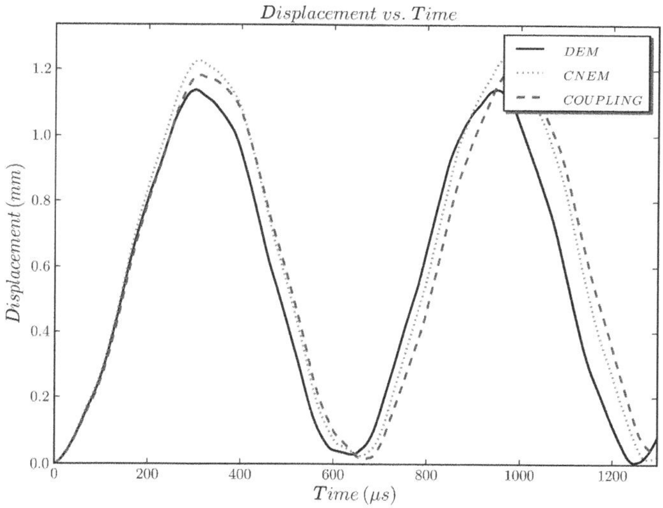

Figure 3.24. *Bending loading: free-end displacements $U_{y\ mean}$ obtained by DEM, CNEM and the coupling method; $L_B = 10\,mm$, coarse mediator space, continuous differentiable weighting and $\varepsilon = 0.05$*

Since the CNEM approach is very close to the FEM method and has practically all its advantages, the performance of the developed coupling approach should not be affected, compared to a DEM-FEM coupling approach. In addition, since CNEM is a meshfree method, the DEM-CNEM coupling approach is better suited to simulate complex highly dynamic problems. With the development of this approach, these problems have become affordable. The second part of this book aims to apply this approach to simulate qualitatively one of the most challenging problems in fast dynamics: laser shock processing of silica glass.

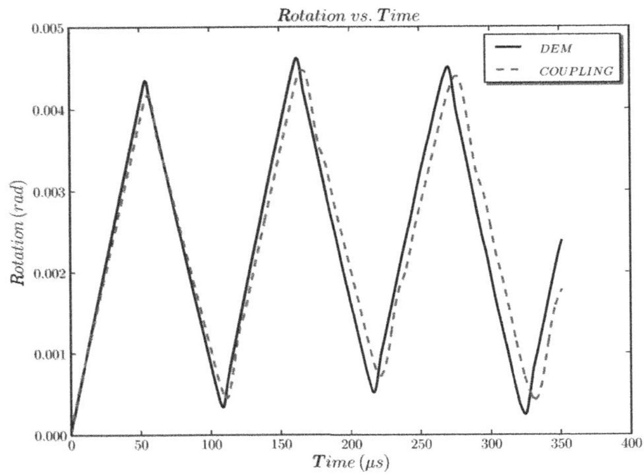

Figure 3.25. *Torsional loading: Free-end rotation $\theta_{x\ mean}$ obtained by DEM, CNEM and the coupling method; $L_B = 10\,mm$, coarse mediator space, continuous differentiable weight functions and $\varepsilon = 0.05$*

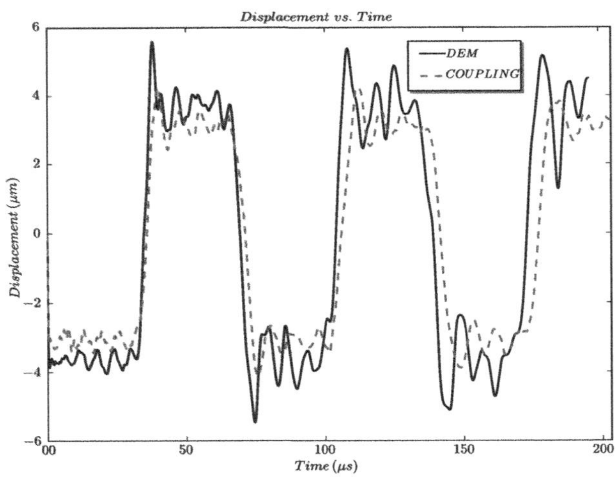

Figure 3.26. *Initial velocity loading: free-end displacement $U_{x\ mean}$ obtained by DEM and the coupling method; $L_B = 10\,mm$, coarse mediator space, continuous differentiable weight functions and $\varepsilon = 0.05$*

PART 2
Application: Simulation of Laser Shock Processing of Silica Glass

4

Some Fundamental Concepts in Laser Shock Processing

4.1. Introduction

In a pioneering piece of work, White [WHI 63] demonstrated in 1963 the ability of high-energy lasers to generate shock waves. His work was extended in 1968 with the development of confined ablation modes [AND 68]. Thereafter, the concept of laser-induced shock waves was used to develop a new material processing technique in the early 1970s, called laser shock processing (LSP). This technique consists of subjecting a target sample to a very short pulse of a high-power laser. Initial studies on LSP were done at the *Battelle* institute (Columbus, Ohio, USA) from about 1968 to 1981. Results of these studies illustrated the beneficial effects of the process for several engineering applications. Although this technique demonstrated its feasibility and potential as an alternative to some conventional material processing techniques, it was abandoned by the *Battelle* researchers after 1981 because of the lack of laser sources to provide real industrial potential. Five years later, new studies supported by the automotive industry and other industrial partners were started in France with the objective of developing a clean, local and efficient LSP process. Considerable attention was paid to the potential industrial applications of the process such as fatigue life improvement, corrosion and wear resistance, and compaction effects on powders. Consequently, a wide variety of efficient high-energy laser devices have been developed. Along with this, LSP has become widely used in various engineering applications to increase the materials resistance to surface-related failures (e.g. fatigue, fretting fatigue and stress corrosion cracking), strengthen thin sections, shape or straighten parts, break up hard materials and even to characterize mechanical behaviors of complex materials (e.g. composites and glasses).

The ever-increasing interest in LSP has created the need for further investigations to properly address some ongoing problems. Damage of the optics constituting the laser devices due to their exposure to high-power lasers is a recurrent problem, resulting in increased process costs and accident risks. Several experimental works have been performed to investigate the origins and sources of the laser-induced damage phenomenon [LES 06, BOU 98, GRA 78, COT 89, FAB 98, KAN 98]. Nevertheless, despite the current state of these works, such a phenomenon is always one of the major concerns encountered in LSP. This is mainly due to the high cost of the laser experiments, as well as the technical difficulties preventing the effectiveness of the experimental work in terms of providing insightful and complete information. Due to the very short duration of the process, much of the useful information cannot be directly measured, observed or is difficult to acquire by current experimental means. Thus, a next generation of research effort is required and the numerical support is necessary to best meet the increasing needs of LSP applications. The second part of the present book aims to put into play the developed discrete-continuum coupling approach to simulate qualitatively the LSP of silica glass which is the dominant constitutional material of the optical equipment in laser devices. The purpose of this qualitative simulation is to ensure that the developed coupling approach can correctly reproduce the important phenomena encountered in LSP. This allows us to illustrate the possibility of using this approach as an inverse technique to provide solutions for several difficulties so far preventing an accurate quantitative simulation of the LSP process.

In order to consider all the important LSP phenomena in this simulation, the present chapter attempts to put forward some fundamental concepts related to LSP of silica glass. First, the theory of laser–matter interaction will be reviewed to better understand the different mechanisms involved in LSP. Then, the silica glass response under high pressures will be detailed to collect the main specificities of its mechanical behavior that will be modeled later using the proposed coupling approach.

4.2. Theory of laser–matter interaction: high pressure generation

The interaction between a laser beam and matter is based on the change of the fast photon energy into thermal and/or kinetic energy in the first atomic layers of the matter. When a material is irradiated with lasers, the laser energy is first transformed into electronic excitation energy and then transferred to lattices of the material through collisions between the electrons and lattices. The deposition of laser energy will produce a series of effects, such as temperature rise, gasification and ionization. The physical processes of laser–matter interaction (Figure 4.1) depend mainly on the laser intensity given by:

$$I_L = \frac{E_L}{\tau_L \, S_L} \tag{4.1}$$

where E_L is the pulse energy, τ_L is the full duration at half maximum (FDHM) and S_L is the energy deposition surface. The present work focuses on laser radiations with laser density above $10^{12}\, W\, cm^{-2}$, leading to plasma generation and ablation of matter.

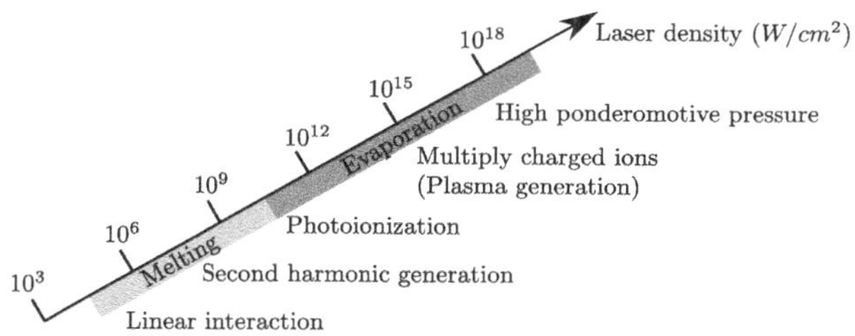

Figure 4.1. *Physical processes of laser–matter interaction*

4.2.1. *Generation of shock wave by laser ablation*

Materials subjected to laser radiations absorb the incident laser energy [HAN 11]. In the case of normal incidence, the coefficient of energy absorption A is given by:

$$A = \frac{4\, n_L}{(n_L + 1)^2 + k_L^2} \tag{4.2}$$

where n_L and k_L are, respectively, the real and imaginary parts of the refraction complex index ($\tilde{n}_L = n_L + i k_L$). Otherwise, the influence of the incidence angle and the polarization of the laser radiation must be taken into account in the computation of A. The laser energy is first absorbed by the free electrons in a small depth l_{skin} determined by:

$$l_{skin} = \frac{\lambda}{2\,\pi\,k_L} \qquad [4.3]$$

with λ being the wavelength of the incident laser wave. Then, the absorbed energy is transferred to the ions by collision over a very short time period τ_{ei}, depending on the optical properties of the irradiated material.

The energy absorption leads, *inter alia*, to material heating. In the case of normal incidence, the temperature raise can be expressed in terms of depth x and laser irradiation time t as follows:

$$\Delta T(x,t) = 2\,(1-R)\,\alpha\,I_L\,\frac{t}{\pi\,k\,\rho\,C}\;\text{ierfc}\left(\frac{x}{2\sqrt{\frac{k\,t}{\rho\,C}}}\right) \qquad [4.4]$$

where R is the reflectivity, α is the absorptivity, I_L is the laser intensity, k is the thermal conductivity, C is the specific heat, ρ is the density of the irradiated material and "erfc" is the inverse complementary error function. According to the FDHM τ_L, two types of laser pulses can be distinguished: short pulses if $\tau_L < \tau_{ei}$ and long pulses if $\tau_L >= \tau_{ei}$. In the case of a short pulse, the energy absorption and the material heating are decoupled (ions are not heated during the laser irradiation). The absorbed energy is transferred to ions only from the free electrons by collision. If the ions energy exceeds the ionization threshold, the heated zone is transformed to a plasma. However, in the case of a long pulse, these phenomena are coupled and the laser radiations interact directly with the plasma. This interaction consists of:

– the temperature of the irradiated material rises and the deposited energy diffuses inside the material to a certain depth (Figure 4.2(a)), causing the material in this region to melt;

– the temperature of melted material sharply increases to over the boiling point due to the heavy deposition of laser energy; nevertheless, the boiling does not start and the liquid is super-heated because of the absence of nucleation (Figure 4.2(b));

– the disturbance will bring about nucleation and the super-heated liquid thickens as the size and the number of bubbles grow (Figure 4.2(c));

– once the size of bubbles is sufficiently large, the startling boiling occurs and afterward the super-heated liquid and particles are ejected (Figure 4.2(d)). This process is known as material laser ablation.

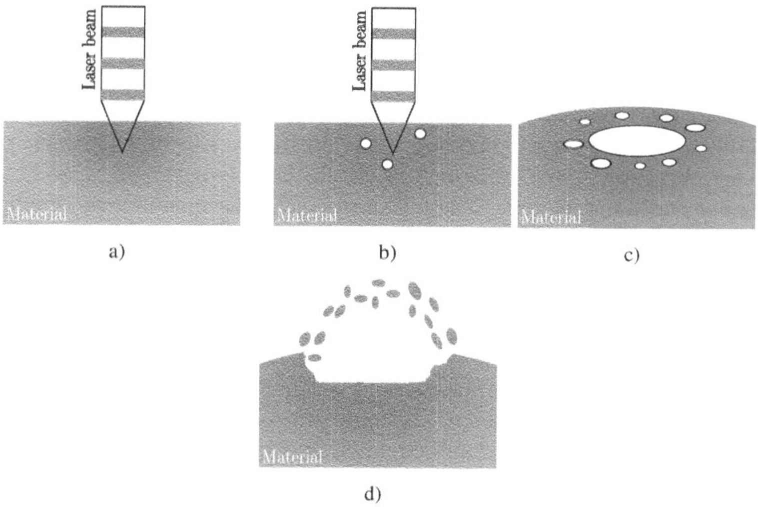

Figure 4.2. *Physical processes of material laser ablation*

As a result of the material laser ablation, high pressures will rise and propagate in the material. The maximum value of the laser-induced pressure P_L^{max} can be determined using experiment-based abaci [LES 06]. In the nanosecond order, P_L^{max} can be approximated using the following empirical formula of Phipps [PHI 02, LES 06]:

$$P_L^{max} = 6.22e^{10} \frac{A^{7/16}}{Z^{9/16}} \frac{(\alpha I_L)^{3/4}}{\left(\lambda \sqrt{\tau_L}\right)^{1/4}} \qquad [4.5]$$

where A and Z are, respectively, the mass and atomic number of the irradiated material. However, it is very difficult to determine experimentally the temporal profile of the laser-induced pressure in materials which is necessary to model any laser application. This can be estimated numerically by solving the equations of laser–matter interaction. Several software packages specialized in such a computation can be found in the literature. Among them, we can cite ESTHER [COL 06] and CHIC [MAI 08]. ESTHER is developed in *CEA/DIF Île-de-France* and allows for one-dimensional (1D) simulations of interaction between light and matter, based on the Helmhotz equations [KHA 05]. Therefore, it can be used to perform computation in the femtosecond regime. CHIC is a bidirectional (two-dimensional (2D)) code developed in the *CELIA* laboratory. The equations of light–matter interaction implemented in this code are based on the theory of collision-induced absorption [FRO 93] which limits its application to the nanosecond regime. These two software packages are the most used to approximate the temporal profile of the laser-induced pressure. However, they are based on numerous simplifying assumptions making them ineffective in the general case. Furthermore, the spatial distribution of the laser-induced pressure is very irregular and unpredictable. These reasons make the quantitative simulation of LSP of materials very challenging. Using an inverse technique to obtain information on the laser-induced pressure from final experimental results would be worth exploring.

This part of laser–matter interaction is not within the scope of this work which aims to simulate qualitatively the mechanical response of silica glass subjected to high-energy lasers. The temporal pressure profile is assumed to be given and will be directly applied on the shock surface (irradiated surface). Also, it is assumed that the thermal effects are relatively slow compared to the propagation of the laser-induced pressures within the material. Therefore, the thermal aspect of LSP will be neglected in the present work.

4.2.2. *Shock wave propagation in materials*

The laser-induced pressure makes the irradiated material under compressive stresses which propagate in all directions [TAN 11, LES 06, FAB 98]. When the amplitude of the stress waves greatly exceeds the dynamic strength of the material, the shear stresses can be neglected in comparison with the hydrostatic component of the stress tensor. We can, therefore, consider a high pressure state traveling into the material which can be assumed, in a first approximation, to have no shear resistance (i.e. the shear modulus is zero). Under this assumption, the material state is completely characterized by three thermodynamic parameters: pressure P, density ρ (or specific volume $v = \rho^{-1}$) and internal energy E (or temperature T). The propagation of a high pressure state into a material can lead to the formation of a shock wave. To simplify the understanding of this concept, an ideal gas will first be considered.

For an ideal gas, the associated equation of state in the case of an isentropic process can be written:

$$PV^{\gamma} = const \qquad [4.6]$$

where V is the volume occupied by the gas, $\gamma = \dfrac{C_p}{C_v}$ is the adiabatic index, C_p and C_v are, respectively, the specific heat at constant pressure and the specific heat at constant volume. Differentiation of equation [4.6] leads to:

$$\frac{\partial P}{\partial V} = -\gamma \frac{P}{V} \qquad [4.7]$$

This last equation shows that the compressibility $K = -\dfrac{1}{V}\left(\dfrac{\partial V}{\partial P}\right)_s$ of an ideal gas decreases with pressure P, since $\dfrac{P}{V}$ increases with P. This is explained physically by the fact that the external electron shells of adjacent atoms penetrate and repel each other. Therefore, the velocity of a disturbance C in an ideal gas, which is equivalent to $\sqrt{\left(\dfrac{\partial P}{\partial \rho}\right)_s} = \sqrt{\left(\dfrac{1}{\rho K}\right)_s}$ in $1D$ configuration, increases with P. In other words, the high amplitude isentropic disturbances travel faster than the low amplitude ones. This is the *sine qua non* requirement for shock wave formation. Therefore, a compression front steepens up as it travels through the material because the higher amplitude regions of the front travel faster than the lower amplitude regions. This leads to a shock wave which is defined simply as a discontinuity in the thermodynamic parameters (P, ρ (or v) and E (or T)). Conversely, a release front spreads out during its propagation through the material (Figure 4.3).

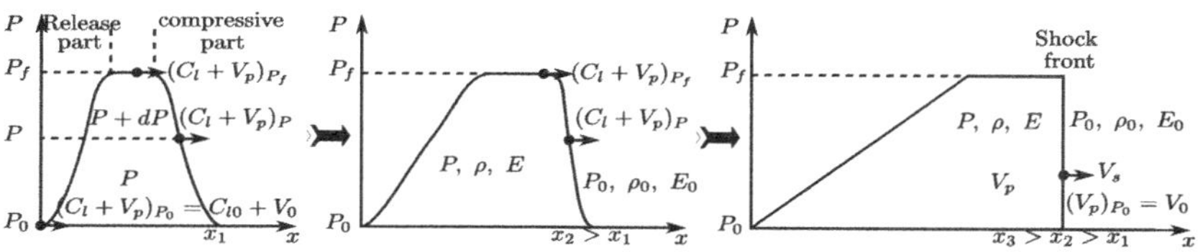

Figure 4.3. *Evolution of a pressure pulse during its propagation in a material*

For a solid, it is necessary to differentiate between the deviatoric and hydrostatic components of stress. When the former is negligible, it becomes possible to apply the treatment for fluids without reservation.

4.2.2.1. *Hydrodynamic treatment*

Shock waves are characterized by a steep front and require a state of uniaxial strain (no considerable lateral flow of material) which allows the buildup of the hydrostatic component of stress to high levels. When the last (hydrostatic component) far exceeds the dynamic flow strength, the material behaves as a fluid. Therefore, the Rankine–Hugoniot conservation equations for fluids [RAN 70, RAN 87, RAN 89] can be applied to calculate the shock wave parameters. This is valid taking into account several assumptions:

– the shock is a discontinuous surface and has no apparent thickness;

– the material behaves as a fluid (no shear resistance); the theory is, therefore, restricted to high pressures;

– body forces (such as gravitational forces) and heat conduction at the shock front are negligible;

– there is no elastoplastic behavior;

– material does not undergo phase transformation.

Taking into account these assumptions, the shock wave equations can easily be obtained by considering a small region Ω_s with cross-section A immediately ahead of and behind the shock front (Figure 4.4).

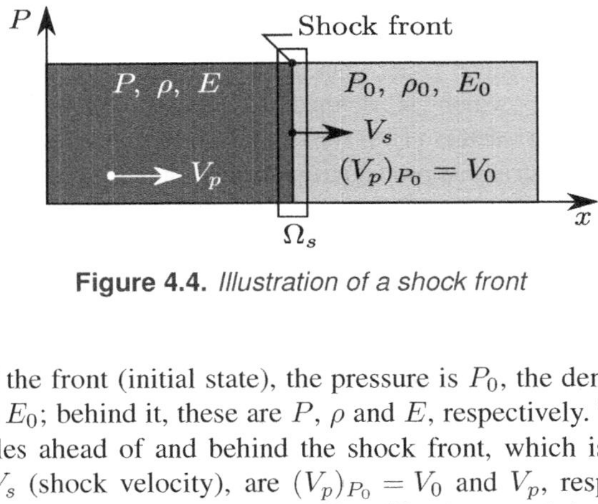

Figure 4.4. *Illustration of a shock front*

Ahead of the front (initial state), the pressure is P_0, the density is ρ_0 and the energy is E_0; behind it, these are P, ρ and E, respectively. The velocities of the particles ahead of and behind the shock front, which is moving at a velocity of V_s (shock velocity), are $(V_p)_{P_0} = V_0$ and V_p, respectively. The apparent velocity of the shock front is $(V_s - V_0)$, because it is moving into a region of particle velocity $V_p = V_0$. At the same time, the material leaving the shock front is moving at a velocity $V_s - V_p$. Using this, the equations of mass, momentum and energy conservation can be derived:

4.2.2.1.1. Mass conservation

This principle states that the mass entering Ω_s equals the mass leaving it:

$$A\,\rho_0\,(V_s - V_0)\,dt = A\,\rho\,(V_s - V_p)\,dt \qquad [4.8]$$

This can be written simply as:

$$\rho_0 \left(V_s - V_0\right) = \rho \left(V_s - V_p\right) \qquad [4.9]$$

4.2.2.1.2. Momentum conservation

The momentum change in Ω_s must be equal to the impulse given to it:

$$\underbrace{\rho A(V_s - V_p)dt}_{mass} \quad \underbrace{V_p}_{Particle\ velocity} \quad \underbrace{-\rho_0 A(V_s - V_0)dt}_{mass} \quad \underbrace{V_0}_{Particle\ velocity} = \underbrace{(P - P_0)\, A\, dt}_{Impulse} \qquad [4.10]$$

Taking into account [4.9], equation [4.10] can be written as:

$$\rho_0 \left(V_s - V_0\right)(V_p - V_0) = P - P_0 \qquad [4.11]$$

4.2.2.1.3. Energy conservation

The conservation of energy postulates that the work of the external forces (pressure in this case) on Ω_s (ΔW) equals the difference in the total energy (kinetic and internal) in Ω_s (ΔE_{tot}):

$$\Delta W = \Delta E_{tot} \qquad [4.12]$$

with:

$$\Delta W = \underbrace{PA}_{Force} \ \underbrace{V_p dt}_{Distance} - \underbrace{P_0 A}_{Force} \ \underbrace{V_0 dt}_{Distance} \qquad [4.13]$$

and

$$\begin{aligned}\Delta E_{tot} = &\frac{1}{2}\left[\rho\, A \left(V_s - V_p\right) dt\right] V_p^2 + E\, A\, \rho \left(V_s - V_p\right) dt \\ &-\frac{1}{2}\left[\rho_0\, A \left(V_s - V_0\right) dt\right] V_0^2 - E_0\, A\, \rho_0 \left(V_s - V_0\right) dt\end{aligned} \qquad [4.14]$$

Taking into account [4.9], equation [4.15] can be obtained:

$$P \left(V_p - V_0\right) = \rho_0 \left(V_s - V_0\right) \left(E - E_0 + \tfrac{1}{2}\left(V_p^2 - V_0^2\right)\right) \qquad [4.15]$$

By summation of [4.11] and [4.15] and introducing [4.9], we can obtain the classic form of the energy conservation:

$$E - E_0 = \tfrac{1}{2}\left(P - P_0\right)\left(v_0 - v\right) \quad \text{with} \quad v = \tfrac{1}{\rho} \qquad [4.16]$$

Equations [4.9], [4.11] and [4.16] are known as the Rankine–Hugoniot equations. These equations involve five parameters which are the pressure P, particle velocity V_p, shock velocity V_s, density ρ and energy E. Therefore, an additional equation is required to determine these parameters as a function of one of them. This fourth equation which can be conveniently expressed as the relationship between the particle and shock velocities is called the equation of state (EOS) and has to be determined experimentally. An empirical polynomial equation is proposed in the literature to describe this relationship [4.17]:

$$V_s = C_0 + S_1\, V_p + S_2\, V_p^2 + \ldots \qquad [4.17]$$

where C_0 is the hydrodynamic sound velocity in the material at zero pressure, expressed in terms of longitudinal C_l and transverse C_t sound velocity as $C_0 = \sqrt{C_l^2 - \frac{4}{3}\, C_t^2}$, and S_1 and S_2 are the empirical parameters. For most materials in the absence of phase transformation, equation [4.17] reduces to a linear relationship ($S_2 = 0$) as shown in Figure 4.5. Several works giving the values of the equation of state (EOS) parameters for different materials can be found in the literature [MEY 81, WAL 55].

4.2.2.2. *Representative relationships for shock waves*

Applying [4.9], [4.11], [4.16] and [4.17], other relationships between the hydrodynamic (P, ρ and E) and kinetic (V_p and V_s) parameters can be determined, e.g. $P - V_s$, $P - V_p$, $P - \rho$ and $E - V_s$ relationships.

4.2.2.2.1. Hugoniot curve

The relationship between P and ρ (or $v = \rho^{-1}$) is usually known as the Hugoniot equation. This equation is defined as the locus of all shocked states in the (P, ρ) or (P, v) frame and essentially describes the material properties (Figure 4.6(a)). The straight line in Figure 4.6(a) relaying the (P_0, v_0) and (P_1, v_1) states is the Rayleigh line and refers to the shock state at P_1. It is very important to realize that when pressure is increased in a shock front, it does not follow the $P - v$ curve. Rather, it jumps discontinuously from P_0 to P_1. The slope of this line is proportional to the square of the shock velocity (V_s).

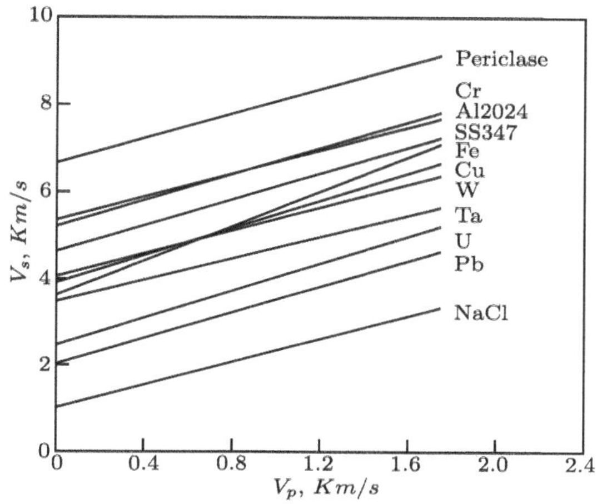

Figure 4.5. *Experimentally measured EOS curve of V_s versus V_p for several materials (taken from [MEY 94])*

4.2.2.2.2. Shock polar

The shock polar (Figure 4.6(b)) is defined as the locus of all shocked states in a material expressed in the (P, V_p) frame. In this plot, the slope of the Rayleigh's line is proportional to the shock velocity (V_s) and is called the shock impedance: $Z = \rho_0 V_s$. In the region of pressures corresponding to low particle velocities in comparison with the shock velocity ($V_p \ll V_s$), the shock impedance is reduced to the acoustic impedance $Z = \rho_0 C_0$. In this case, the shock polar can be approximated by a straight line.

4.2.2.3. *Permanent deformation and shock wave*

In the above shock wave description, the shock front is assumed to be a state of hydrostatic pressure P, for which the corresponding stress tensor can be expressed in a Cartesian coordinate system as follows:

$$\boldsymbol{\sigma} = \begin{pmatrix} P & 0 & 0 \\ 0 & P & 0 \\ 0 & 0 & P \end{pmatrix} \qquad [4.18]$$

This is particularly true in the case of pressures widely exceeding the dynamic material strength. In the region of low shock pressures, the irreversible effects which raise during the shock wave propagation become significant. These effects are illustrated, hereafter, using a simplified elastoplastic model based on the Tresca criterion.

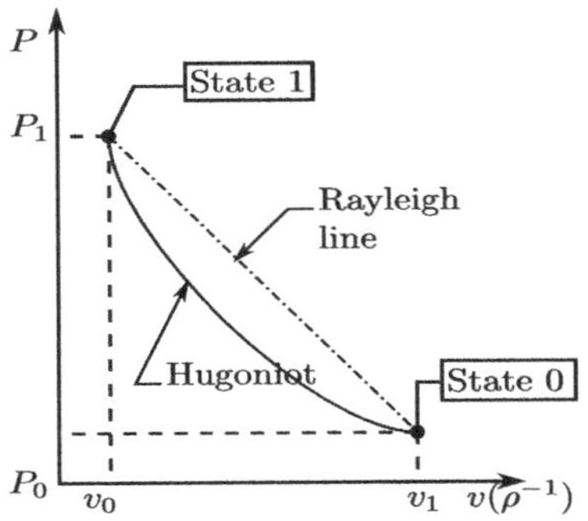

a) Hugoniot curve: Pressure (P) *vs.* specific volume (v)

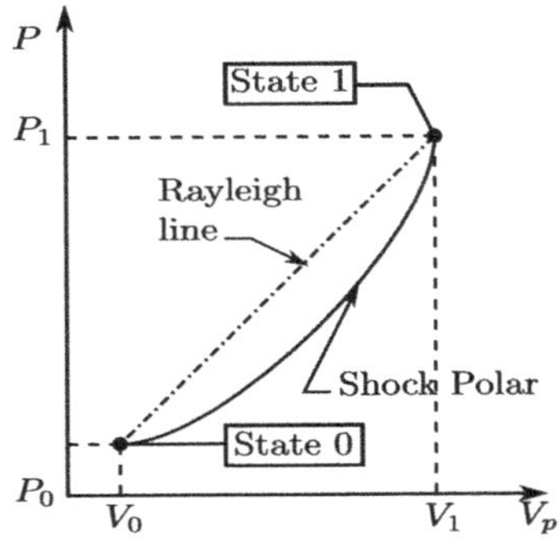

b) Shock polar: Pressure (P) *vs.* particle velocity (V_p)

Figure 4.6. *Representative plots for shock waves*

When a monodimensional shock wave is traveling inside a homogeneous and isotropic solid, a state of uniaxial strain is established. Under these conditions, the deviatoric component of the stress tensor is non-zero. Assuming that the shock wave is traveling in the x direction, the stain and stress tensors are, respectively, written as:

$$\varepsilon = \begin{pmatrix} \varepsilon_x & 0 & 0 \\ 0 & 0 & 0 \\ 0 & 0 & 0 \end{pmatrix} \qquad [4.19]$$

$$\begin{aligned} \boldsymbol{\sigma} &= \qquad \boldsymbol{\sigma}_s \qquad + \qquad \boldsymbol{\sigma}_d \\ &= \begin{pmatrix} K\varepsilon_x & 0 & 0 \\ 0 & K\varepsilon_x & 0 \\ 0 & 0 & K\varepsilon_x \end{pmatrix} + \begin{pmatrix} \frac{4}{3}G\varepsilon_x & 0 & 0 \\ 0 & -\frac{2}{3}G\varepsilon_x & 0 \\ 0 & 0 & -\frac{2}{3}G\varepsilon_x \end{pmatrix} \end{aligned} \qquad [4.20]$$

where K and G are, respectively, bulk and shear moduli of the material. Under the dynamic elastic limit, called the Hugoniot elastic limit Hugoniot elastic limit (HEL), the solid behaves in a perfectly elastic manner. Beyond this value, it begins to exhibit signs of permanent deformation (plasticity). In terms of material strength and using the Tresca criterion, a material deforms plastically when the shear stress τ_{max} equals the yield shear stress τ_0 :

$$\tau_{max} = \frac{1}{2} \max_{i,j} |\sigma_i - \sigma_j| = \tau_0 \qquad [4.21]$$

where σ_i, $i = 1, 2, 3$ are the principle stresses. Using [4.21], the relationship between HEL and τ_{max} can be derived:

$$\begin{aligned} P &= \frac{1}{3}(\sigma_1 + \sigma_2 + \sigma_3) \\ &= \frac{1}{3}(\sigma_x + 2\sigma_y) \\ &= \sigma_x - \frac{2}{3}(\sigma_x - \sigma_y) \end{aligned}$$

or :

$$\tau_{max} = \frac{1}{2}(\sigma_x - \sigma_y) \qquad [4.22]$$

thus,

$$\sigma_x = P + \frac{4}{3}\tau_{max} \quad \Longrightarrow \quad HEL = P_{HEL} + \frac{4}{3}\tau_0 \qquad [4.23]$$

When the HEL is reached, the Hugoniot curve shows a change in the slope (Figure 4.7). Furthermore, the elastic portion of the wave is separated from the plastic wave (Figure 4.7). We can distinguish two shock fronts: elastic shock front traveling at the longitudinal sound velocity C_l and plastic shock front traveling at the shock velocity V_s. The consequences of this phenomenon will be discussed in the next section (4.2.2.4). At sufficiently high pressures, the plastic shock front overcomes the elastic shock front, since V_s increases with P. This justifies the application of the hydrodynamic theory of shock wave in the region of high pressures.

4.2.2.4. *Phenomena spoiling the propagation of shock wave*

4.2.2.4.1. Hydrodynamic damping

For most materials, the concavity of the Hugoniot curve (Figure 4.6(a)) is positive. Therefore, the compression front steepens as it travels through the material because the higher amplitude regions of the front travel faster than the lower amplitude regions. On the contrary, the release front spreads out (the release rate decreases) as it travels through the material. The head of the release part (at the maximum pressure) is traveling at a velocity $(V_p + C_l)_{P_f}$, where V_p and C_l are the particle velocity and the longitudinal sound velocity at the pressure P_f. Since $(V_p + C_l)_{P_f} > V_s$, the release part of the wave overtakes the front as the wave progresses. This progressively reduces the pulse duration to zero (Figure 4.8). After it is zero, the peak pressure and so the shock velocity start to decrease (Figure 4.8). This is generally referred to as "hydrodynamic damping".

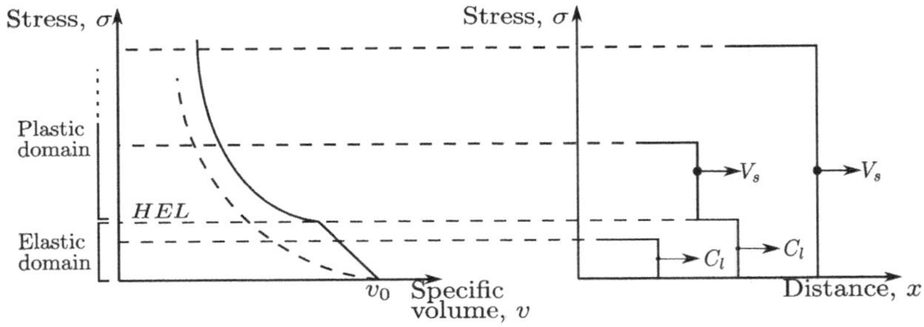

Figure 4.7. *Permanent deformation effects on the Hugoniot and the wave propagation; left: Hugoniot curve (solid line), isentropic hydrostatic compression (dashed line); right: shock wave instability due to the HEL presence (dual wave)*

4.2.2.4.2. Elastoplastic damping

In addition to the hydrodynamic damping, the shock waves in solids can also be attenuated due to the difference between the velocity of the elastic and plastic fronts. As seen previously, beyond the HEL, the shock wave is split into two parts: elastic shock front and plastic shock front traveling, respectively, at velocities C_l and V_s. During rarefaction (release), the elastic part of the release wave follows the shock front at velocity C_l which is, in the region of low pressures, greater than V_s. In this case, it can overtake the shock front and reduce the peak pressure (Figure 4.8). This mechanism is classically known as "elastoplatic damping".

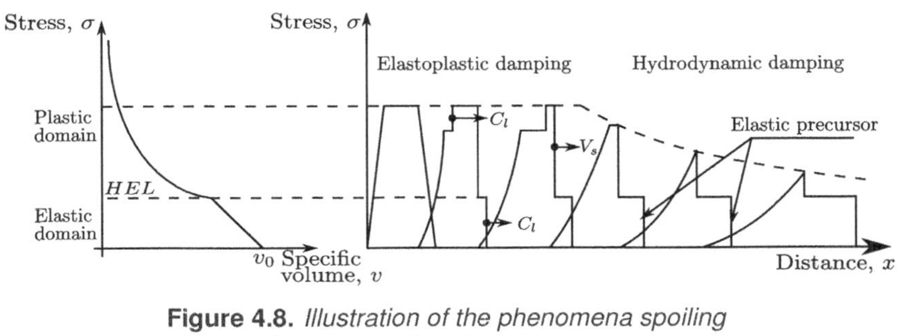

Figure 4.8. *Illustration of the phenomena spoiling the propagation of shock waves*

4.2.2.5. *Shock wave transmission and reflection*

During its propagation, the shock wave can migrate between media, e.g. when reaching the free surface (material/air interface) in the case of LSP processing, which can result in modification of thermodynamic and kinetic parameters. This depends mainly on the shock impedance of the considered media ($Z = \rho_0 V_s$). The shock impedance of materials increases as the sonic velocity and density increase. The best way to study the wave transfer between two media is by means of the impedance matching technique. The continuity at the interface between the materials dictates that the particle velocity and pressure must be the same in both the interface sides. Considering the case of a shock wave of pressure P_1 propagating between two media A and B having, respectively, impedance Z_A and Z_B, two cases can be distinguished.

– $Z_A < Z_B$: the incident pressure P_1 changes at the interface to ensure the dynamic equilibrium in this region. The interface pressure can graphically be determined using the shock polars of medium A and B. Plotting these curves (Figure 4.10), the symmetric curve AR of the curve A, with respect to the pressure axis, passing through the shock pressure P_1 can be plotted. This curve intersects the curve B at P_2. This is the pressure in the medium B. As the shock front reaches the interface, the pressure rises to P_2 and propagates into both A and B. The bottom view of Figure 4.10 presents the associated sequences of pressure profiles. The same reasoning applies to determine the particle velocity in both A and B at the interface V_{P_2}.

– $Z_A > Z_B$: similar reasoning as in the previous case can be applied while reversing the curves A and B (Figure 4.9). In this case, P_2 is lower than P_1 ($P_2 < P_1$) which produces a release pulse to be sent through the medium A. It should be recalled that the release portion of the wave is not a shock front, but it has a slope that decreases as the propagation velocity increases.

The second case ($Z_A > Z_B$) is frequently encountered in LSP experiments, which explains the spalling phenomenon often observed at the rear surface of the irradiated material. When the laser-induced shock wave reaches this surface (in general material/air interface), a release pulse is produced and sent through the irradiated material. When this pulse crosses the incoming release portion due to the fast laser unloading (at t_4 in Figure 4.9), a tensile state of high level takes place in the crossing region. If the resulting tensile pulse is sufficiently powerful, material spalling can occur. This phenomenon will be studied in Chapter 6.

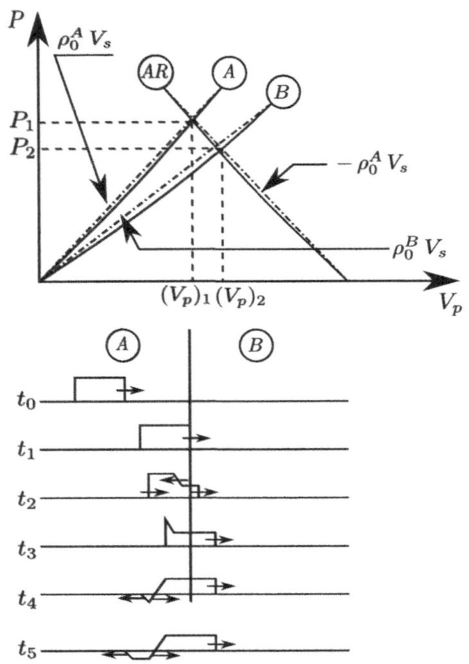

Figure 4.9. *Transmission of shock wave from material A with high impedance to material B with low impedance*

4.2.3. *Laser-induced damage in materials*

When a material is irradiated by a laser beam, there are mainly three types of processes that can cause material damage [WOO 03]:

– thermal processes: these arise from absorption of the laser energy in the material, and generally occur for continuous wave operation, long pulse lengths and high-pulse-repetition-frequency pulse trains;

– dielectric processes: these arise when the electric field density is high enough to strip electrons from the lattice. Such processes take place when the pulse lengths are short enough for avalanche ionization to take place and when the thermal absorption is low enough for the avalanche threshold to be below the thermal threshold;

– mechanical processes: by propagation of a non-sustained shock wave into a solid, it is possible to produce strong levels of tensile leading to damage the material [ANT 02]. The laser-driven shock wave propagates into the material and reverberates on the back surface (free surface) into a release wave which interacts with the incoming release wave due to the fast laser unloading. The crossing of these two release waves results in tensile state inside the material. The distance between the tensile zone and the back surface is proportional to the pulse duration. If the amplitude and the duration of the tensile state are sufficient, there can be fracture of the considered material and formation of spalls (Figure 4.11(a)). The incident wave continues to reflect on the new free surfaces (generated by spalling) and the fracture threshold σ_f can be reached other times. Therefore, further spalls can be formed (Figure 4.11(b)). The spalling process continues as long as the incident wave is sufficiently intense. This is known as multiple spalling. Each spall is generated from a lower compression state, and thus with a lower initial velocity. The multiple spalling is clearly evidenced in Figure 4.12 taken from the experimental work of Rinehart [RIN 52].

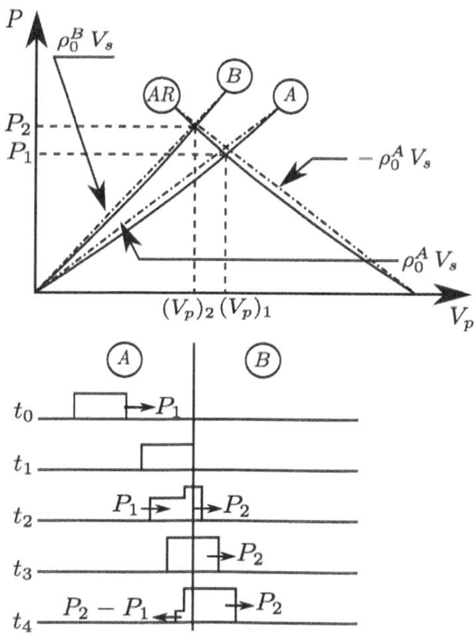

Figure 4.10. *Transmission of shock wave from material A with low impedance to material B with high impedance*

The first two processes are due to laser energy absorption (first step of LSP: laser–matter interaction) and generally occur in a small depth from the front face of the irradiated material. These processes are not within the scope of this book which focuses on the laser-induced damage from a mechanical point of view. Therefore, this work focuses only on the third process which will be studied numerically in Chapter 6.

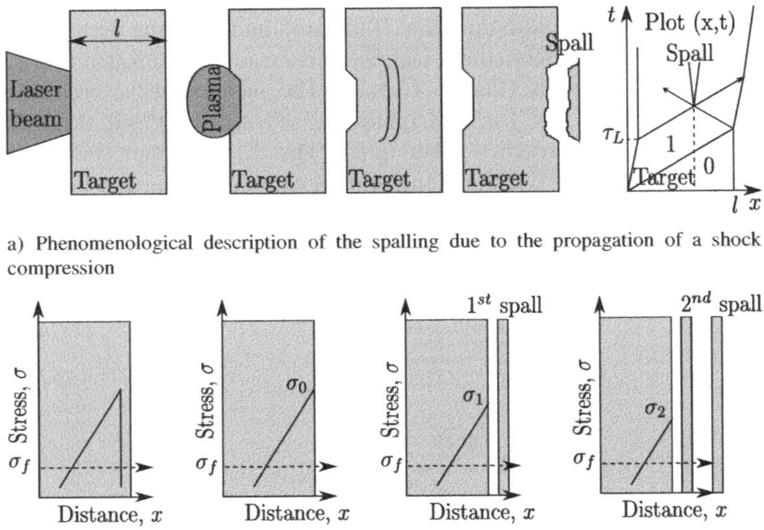

a) Phenomenological description of the spalling due to the propagation of a shock compression

b) Multiple spalling process

Figure 4.11. *Laser-induced spalling*

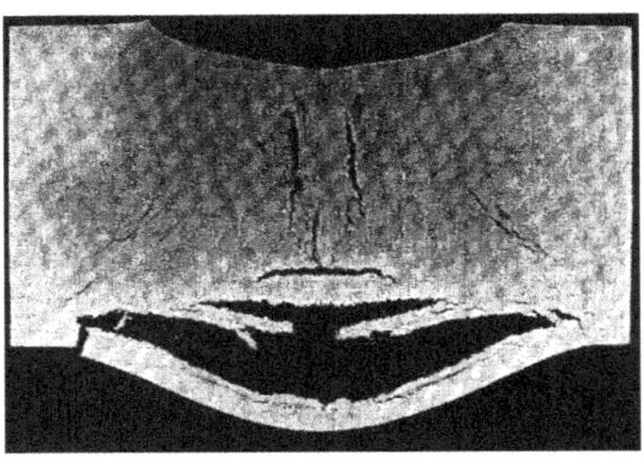

Figure 4.12. *Example of a target that has undergone multiple spalling (taken from [RIN 52])*

4.3. Mechanical response of silica glass under high pressure

According to the dependence of various properties such as density, hardness, refractive index, Young's and shear moduli with the pressure and fictive temperature (defined in a simple manner as the temperature at which the glass would be in thermal equilibrium), glasses can be classified under three headings: normal, intermediate and anomalous. Normal glasses, such as soda-lime silicate glass [LE 03, ROU 00], exhibit properties decreasing with increasing fictive temperature and pressure. Anomalous glasses, such as germania, exhibit the opposite trend in these properties. Intermediate between normal and anomalous glasses are expected to have some properties that are independent of these parameters (i.e. pressure and fictive temperature). Silica glass, which is a typical amorphous material, belongs to the class of anomalous glasses. This material presents an anomalous behavior in its thermal and mechanical properties [BRÜ 70, BRÜ 71] which generally increase with pressure and fictive temperature. This section will detail the main specificities of the silica glass mechanical behavior. To better understand this behavior, the silica glass response under static hydrostatic compression will be first studied. And after, the particularity of this behavior under shock compression will be discussed.

4.3.1. *Silica glass response under quasi-static hydrostatic compression*

4.3.1.1. *Elasticity*

In the region of ordinary pressures (up to approximately $8\,GPa$), the silica glass behaves in a perfectly elastic manner. As mentioned in several experimental works [BRI 48, KON 81], for relatively low pressures (under $3\,GPa$), this material shows a nonlinear elastic behavior which can be explained by the pressure dependence of the bulk and shear moduli, denoted by, respectively, K and G. The experimental works of Kondo *et al.* [KON 81] have demonstrated that these moduli decrease with increasing pressure up to $2.4 - 2.5\,GPa$, after which they increase with increasing pressure until $P_{nle} = 3\,GPa$. Figure 4.13 which is taken from reference [KON 81] shows the pressure dependence of the elastic moduli. Kondo *et al.* [KON 81] have also reported that the experimental results associated with this nonlinear behavior are reproducible. It would, therefore, seem that there is no observable permanent change in these moduli and this behavior is perfectly elastic.

Beyond approximately $P_{nle} \approx 3\,GPa$, the silica glass behavior returns linear elastic until the densification pressure $P_d \approx 8\,GPa$, where the silica glass begins to exhibit signs of permanent deformation. Table 4.1 presents the initial elastic properties of silica glass which are used in the present work.

Young's modulus	*Poisson's ratio*	*Density*
$E = 72\,GPa$	$\nu = 0.17$	$\rho = 2200\,Kg\,m^{-3}$

Table 4.1. *Elastic properties of silica glass*

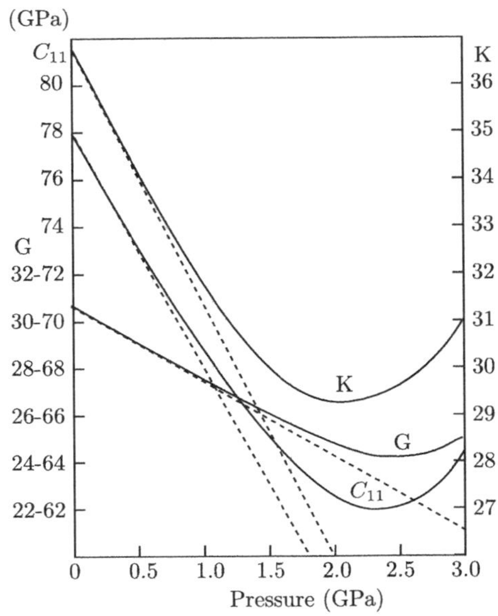

Figure 4.13. *Relationships between the elastic moduli of silica glass and the hydrostatic pressure; K, G and C_{11} are, respectively, the bulk, shear and longitudinal moduli (taken from [KON 81])*

4.3.1.2. *Permanent deformation: densification*

Contrary to crystalline solids, when silica glass (amorphous material) is subjected to high hydrostatic compression (in the range of $8\,GPa$ or beyond), it begins to exhibit signs of permanent deformation which is quite different from the plastic flow observed in crystalline solids. Indeed, the plastic flow in crystalline solids is volume-conservative and initiates under shear stress only. The spherical part of the stress tensor has no influence on the plastic flow. However, the permanent deformation of silica glass is always accompanied by a volume change. In addition, it can initiate under hydrostatic pressure as reported in several previous works [BRI 53, KER 08, ROU 08]. For these reasons, the permanent deformation of silica glass is generally called "densification" (and not plasticity like in crystalline solids). This term is defined as a volume change ($\Delta V/V_0$) or density change ($\Delta\rho/\rho_0$). These quantities are linked through the equation of mass conservation which can be written as:

$$\frac{\Delta V}{V_0} = -\frac{\dfrac{\Delta\rho}{\rho_0}}{1 + \dfrac{\Delta\rho}{\rho_0}} \qquad [4.24]$$

The densification of silica glass which was discovered in the 1950s has been widely studied in the literature [BRÜ 70, BRÜ 71, BRI 53, KER 08, ROU 10, JI 07]. This phenomenon, which initiates at the densification pressure $P_d \approx 8\,GPa$, continues to evolve until the saturation pressure $P_s \approx 20\,GPa$ and can reach 21 % of density change at this pressure. During the densification process, the mechanical properties of silica glass show an anomalous dependence on the pressure [BRÜ 70, BRÜ 71, BRI 53, KER 08, ROU 10]. Indeed, and contrary to normal glasses, these properties increase with increasing pressure beyond P_d. According to [JI 07, KER 08], Young's modulus and Poisson's ratio of silica glass present, respectively, a spectacular augmentation of 46 % and 56 % at the end of the densification process (Figure 4.14). Beyond the saturation pressure P_s, the densification process stops evolving and the material regains its elastic behavior.

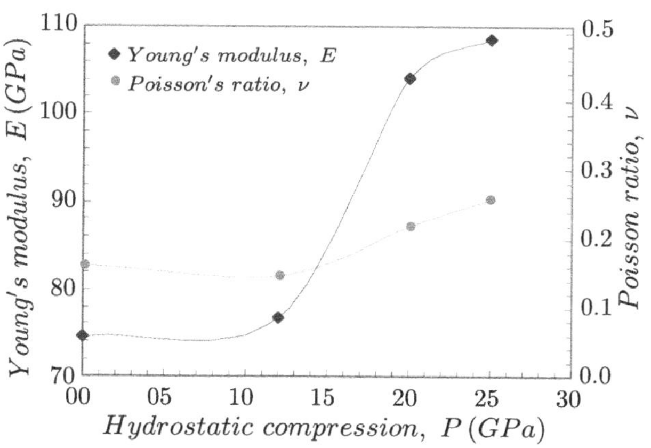

Figure 4.14. *Variation of the silica glass mechanical properties with densification (experimental works of Ji [JI 07])*

NOTE.– Several works studying the silica glass densification can be found in the literature [BRÜ 70, BRÜ 71, KER 08, BRI 53, JI 07, POL 93, ZHA 94, SUG 81, SUG 96, MAR 80, WAC 62, LAL 91]. These works give different values of the densification parameters, such as P_{nle}, P_d, P_s, etc. (Table 4.2). Furthermore, it has been shown that these parameters depend hugely on the type of the samples used to study the densification [COH 61, CHR 62, BRI 53, JI 07]. For example, the results obtained using powder samples are very different from those obtained using massive samples (Figure 4.15). The different values presented above, which were obtained by Ji [JI 07] using thick samples, are for illustrative purposes. In the present work, the silica glass static behavior will be studied numerically using the densification parameters obtained by Ji [JI 07].

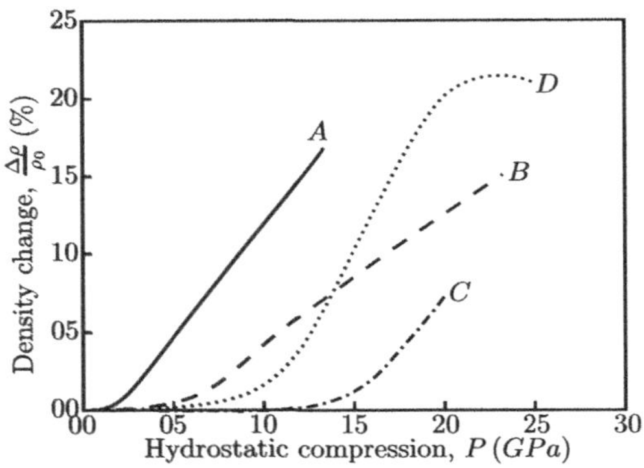

Figure 4.15. *Evolution of silica glass densification with pressure using different samples (ex-situ measures); A: silica powder [COH 61], B: thin disks [CHR 62], C thin disks [BRI 53] and D: thick samples [JI 07] (taken from [JI 07])*

4.3.1.3. *Shear stress effects on the silica glass densification*

Almost all the works on the densification of silica glass agree that the densification process can take place under hydrostatic pressure. However, the shearing effects on this phenomenon present a central issue and need more investigations. Several works on the subject can be found in the literature [ROU 10, CHR 62, MAC 63], some of which are controversial. It began with the work of Bridgman and imon [BRI 53]. They demonstrated that amorphous silica glass exhibits permanent densification which translates into permanent volume decrease of approximately $7.5\,\%$ at around $20\,GPa$. They also demonstrated that this propensity decreases when other substances, such as sodium oxide (Na_2O), are added. However, due to technical difficulties encountered at that time to generate strictly hydrostatic conditions above $5\,GPa$ [BRI 53], uniaxial pressures were applied on thin disks to study the silica glass densification. Unfortunately, these tests induced additional shear stresses, which explain the systematic fracturing of the studied samples. Cohen and Roy [COH 61] studied the densification of silica powder under uniaxial pressures. They found that the densification begins at approximately $2\,GPa$ ($10\,GPa$ in Bridgman and imon's work [BRI 53]) and the volume reduction can far exceed $7.5\,\%$. According to [WEI 62], the difference between the previous results is due to the shearing effects, which are most important in the case of silica powder and allow for greater densification levels to be reached. This explanation was belied later by Cohen and Roy [COH 62], who excluded the shearing effects on the densification level. For more details, the readers can refer to [KER 08]. Despite the numerous works on this subject, the role of the shearing stresses on the silica glass densification remains at present ambiguous. Chapter 5 will try to numerically corroborate the experimental works on the subject.

	$P_d\,(GPa)$	$P_s\,(GPa)$	$\frac{\Delta V}{V_0}$ (%)
Quasi-static studies [BRÜ 70, BRÜ 71, KER 08, BRI 53, JI 07]	$2-10$	$20-30$	$7.5-17.4$
Dynamic studies [POL 93, ZHA 94, SUG 81, SUG 96, MAR 80, WAC 62, LAL 91]	$8.8-10$	$16.6-30$	41

Table 4.2. *Dispersion range of the silica glass densification parameters*

4.3.1.4. *Fracture*

It is known that the silica glass is the most common example of amorphous brittle materials. The fracture of this material is an especially challenging problem [WIE 04]. It depends on several parameters such as the loading, microscopic defects and flaws existing in the material. Also, it is demonstrated that this material can interact with its environment to modify the fracture properties [MIC 83, WIE 67]. These properties must be separated between "slow crack growth" and sudden fracture. In the former, the environment, especially water, interacts with strained bonds at the tip of existing cracks and can induce physi- and chemi-sorption processes which weaken the bond and allow the crack to advance slowly. In the latter, the crack is assumed to propagate without interaction with the environment. For the sake of simplicity, the present work only considers the case of sudden fracture in which the failure process is totally determined by mechanical considerations. As will be seen in section 4.3.2.3, this simplifying assumption might be used without remorse to study the spalling phenomenon which occurs inside the material where the flaw density is very weak.

According to [LAW 93], brittle fracture is thought to be initiated under tensile stress in mode I. The theoretical brittle fracture strength σ_f^{th} of a material without preexisting flaws can be calculated based on its structure and interatomic bond properties. This value is known as the cohesive strength of the material. It can be calculated for a given material by knowing its Young's modulus E, surface energy γ, atomic structure and interatomic potential [ORO 49]. Using the Morse interatomic potential, σ_f^{th} can be expressed as:

$$\sigma_f^{th} = \sqrt{\frac{E\,\gamma}{4\,r_0}} \qquad [4.25]$$

where r_0 is the interatomic distance. Applying [4.25], the silica glass cohesive strength can be obtained: $\sigma_f^{th} = 23\,GPa$. This value widely exceeds the measured strength σ_f^{exp} which is about a few tens of MPa. The discrepancy between the theoretical and measured strength is due to the surface flaws. To correct this problem, Griffith [GRI 21] proposed another equation to calculate the fracture strength for a material containing a crack of length c:

$$\sigma_f^{G} = \sqrt{\frac{2\,E\,\gamma}{\pi\,c}} \qquad [4.26]$$

The result obtained using this last equation [4.26] compares favorably with the measured one, provided that the crack length c is correctly chosen. However, the choice of this parameter c is not a straightforward issue and represents, at present, an open question. In practice, such a parameter is determined experimentally, but the great dispersion of the experimental results makes this way ineffective.

4.3.2. *Silica glass response under shock compression*

In statics, the silica glass mechanical behavior can be divided into three domains: elastic domain between 0 and $8\,GPa$; densification domain between $8\,GPa$ and $20\,GPa$; saturation domain beyond $20\,GPa$. Also, it is shown that the elastic domain of this material can be divided into two parts: nonlinear elasticity up to $3\,GPa$ and linear elasticity between 3 and $8\,GPa$.

In dynamics, these different phenomena have also been reported by several researchers [SUG 81, SUG 96, MAR 80, WAC 62, LAL 91]. However, the densification parameters obtained in this regime are different from those obtained in quasi-statics (Table 4.2). The discrepancies between the quasi-static and dynamic results are not well explained in the literature. They can be due to dynamic effects of silica glass or simply due to results dispersion as encountered in quasi-statics. Especially, analyzing the dynamic studies of silica glass densification, the first point that can be noted is the great dispersion of the experimental results which differ from an experimental work to another. For example, the HEL (P_d in the case of hydrostatic pressure) varies from $8\,GPa$ [SUG 81] to $10\,GPa$ [LAL 91]. Table 4.2 summarizes the dispersion of the densification parameters in dynamics. In the present work, the silica glass dynamic behavior will be studied numerically using the densification parameters obtained by Marsh [MAR 80]. Hereafter, the effects of the silica glass dynamic behavior on the propagation of shock compression will be reviewed.

4.3.2.1. *Propagation of compression wave in the region of reversibility*

As a result of the nonlinear elastic behavior under approximately $3\,GPa$, the wave propagation velocity decreases with increasing particle velocity up to approximately $300\,ms^{-1}$ [WAC 62, SUG 81]. This leads to the formation of an elastic precursor which spreads out as the wave propagates into the material. Consequently, in the case of low stresses (under $3\,GPa$), it is impossible to generate a shock wave in silica glass as reported in several experimental works (Figure 4.16). Beyond this value, since the silica behavior returns linear elastic, the compression front will steepen up as the wave propagates into the material and a shock wave will be generated. Figure 4.16 shows the particle velocity profiles for different compression levels (taken from [SUG 81]). As can be seen in this figure, there is a sudden jump of particle velocity beyond $3\,GPa$ which indicates the formation of a shock wave.

4.3.2.2. *Propagation of compression wave in the region of irreversibly*

Another peculiarity of the silica glass response under shock is the formation of a dual wave in the region of high pressures due to the change in the slope of the Hugoniot curve as mentioned in almost all the previous works studying the silica glass dynamic behavior. The dual wave in silica glass is very similar to that obtained in the case of elastoplastic solids: the shock front splits into reversible (elastic) compression wave and irreversible (densification) compression wave (Figure 4.16).

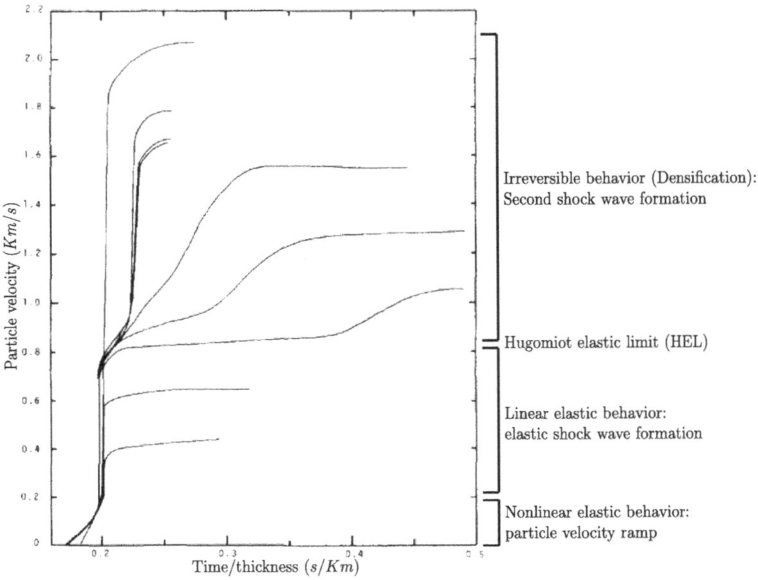

Figure 4.16. *Particle velocity profiles in silica glass under shock compression: nonlinear elastic behavior, linear elastic behavior and apparition of a second wave beyond the HEL (taken from [SUG 81])*

Beyond the HEL, silica glass exhibits a gradual acceleration of the particle velocity [LAL 91]. This can be explained by the augmentation of the bulk modulus K in the densification region. According to [LAL 91], this modulus K can reach $400\,GPa$ at the end of the densification process (Figure 4.17). Figures 4.18 and 4.19 present, respectively, the Hugoniot curve and shock polar of silica glass under shock compression.

4.3.2.3. *Spalling phenomenon*

When a non-sustained shock wave propagates into a solid (as generally encountered in LSP), it is possible to produce high levels of tensile state inside it, which can lead to solid spalling. The first point to note is that the spalling fracture strength is generally much greater (in the range of GPa) than the fracture strength measured in quasi-static tests (a few tens of MPa). This can be explained by the fact that the spalling process occurs inside the material where the flaws density is very weak. This process is, therefore, generated through the internal texture of the material. On the contrary, in the case of quasi-static test, the fracture results from the propagation of the preexisting surface flaws which requires lower energy to be activated.

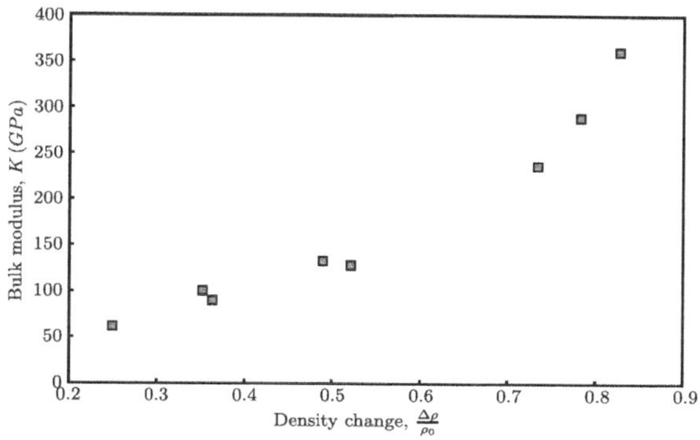

Figure 4.17. *Evolution of the silica glass bulk modulus with densification (taken from [LAL 91])*

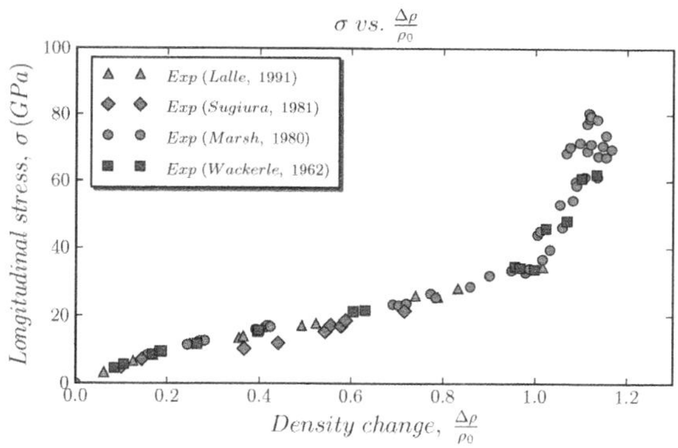

Figure 4.18. *Hugoniot curve of silica glass obtained experimentally (taken from [MIC 07])*

As reported by de Resseguier [DE 92], the spalling strength depends on the loading velocity, the amplitude and duration of the induced tension and the distance between the tensile zone and free surface. The smaller this distance, the smaller the spalling strength is (activation of the surface defects). It should be noted that this parameter is very difficult to be experimentally determined. This is attributed to the existence of "failure waves" [BOU 98] that propagate immediately behind the shock front and damage the material before the spalling process can take place. Depending on the glass type, this parameter varies from 500 MPa for borosilicate glass to more than 2.5 GPa for silica glass [MIC 07].

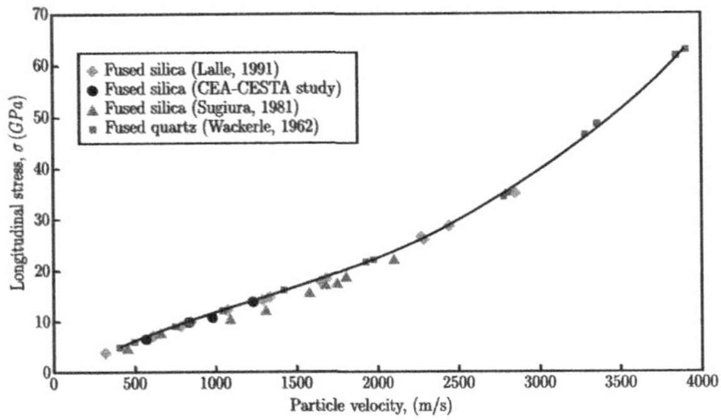

Figure 4.19. *Shock polar of silica glass obtained experimentally (taken from [LAL 91])*

4.3.3. *Summary of the silica glass response under high pressure*

As shown in Table 4.2, different values of the silica glass parameters (P_{nle}, P_d, P_s, etc.) are given in the literature [BRÜ 70, BRÜ 71, KER 08, BRI 53, JI 07, POL 93, ZHA 94, SUG 81, SUG 96, MAR 80, WAC 62, LAL 91]. To develop the new silica glass model, Ji's results [JI 07] are used for quasi-static studies. These results are the most recent (accurate equipments would have been used to obtain them). Furthermore, the author [JI 07] led a complete experimental study from the investigation of silica glass behavior to the Vickers microindentation using samples manufactured in the same conditions. This would reduce the experimental errors due to different manufacturing conditions of the samples. For dynamic studies, the Marsh's [MAR 80] results are used. This author has carried out an interesting experimental work involving pressures of up to $80\,GPa$. Furthermore, his results are in good agreement with other results obtained by Michel [MIC 07], Wackerle [WAC 62], Sugiura [SUG 81] and Lalle [LAL 91]. The main specificities of the silica glass mechanical behavior and the associated numerical values which are considered in the present book can be summarized as follows:

– under quasi-static hydrostatic compression:

- nonlinear elastic behavior up to $P_{nle} = 3\,GPa$ due to the pressure dependence of the bulk and shear moduli,

- linear elastic behavior between P_{nle} and the $P_d = 8\,GPa$,

- densification behavior between the P_d and $P_s = 20\,GPa$,

- Young's and shear moduli increase, respectively, by $46\,\%$ and $56\,\%$ at the end of densification (*ex-situ* measures),

- beyond P_s, the saturation domain takes place,

- fracture strength is about a few tens of MPa. In this work, it is fixed at $\sigma_f = 50\,MPa$;

– under shock compression:

- nonlinear elastic behavior up to $P_{nle} = 3\,GPa$ due to the pressure dependence of the bulk and shear moduli,

- linear elastic behavior between P_{nle} and the $HEL = 10\,GPa$,

- densification behavior between the HEL and $30\,GPa$,

- increase in bulk modulus with densification to $400\,GPa$ at the end of densification process,

- beyond $30\,GPa$, the saturation domain takes place,

- spalling threshold of silica glass is higher than $2.5\,GPa$. For the qualitative in Chapter 6, this parameter is fixed at $2.5\,GPa$.

4.4. Conclusion

The present chapter reviewed the main scientific advancements in the different disciplines that interact to accomplish the second goal of this book: qualitative study of the LSP of silica glass.

First, some basic concepts related to the theory of laser–matter interaction were given in section 4.2. The physics behind this theory can be divided into two steps: shock wave generation by laser ablation and shock wave propagation within the irradiated material. The first step is not within the scope of this work which only focuses on the mechanical part of the LSP of silica glass. The temporal pressure profile generated by the laser–matter interaction (in the first step) is assumed to be given and will be directly applied on the shock surface (irradiated surface). Furthermore, the laser-induced thermal effects generated in a small near-surface region are assumed to propagate slowly compared to the shock wave propagation. Therefore, these effects are not considered in this work. During its propagation within a material, the generated shock wave is subjected to several mechanical phenomena such as hydrodynamic and elastoplastic damping, reflection and transmission (when migrating between media). If the amplitude of the shock wave is high enough, the material in which this wave propagates can also be damaged. These different phenomena will be numerically studied in Chapter 6.

In order to better characterize the silica glass response under high pressures generated by laser shock, section 4.3 reviewed the main experimental works on the complex mechanical behavior of this material. As shown, silica glass belongs to the category of anomalous glasses. It shows an anomalous behavior in its thermal and mechanical properties. In the region of ordinary pressures, it behaves in a perfectly elastic manner up to the densification pressure P_d. The silica glass elastic behavior is divided into two parts: nonlinear elasticity until P_{nle}, and then linear elasticity between P_{nle} and P_d. Beyond P_d, silica glass begins to exhibit signs of permanent deformation called densification up to the saturation pressure P_s. After P_s, this material saturates and returns elastic. During densification, the silica glass mechanical properties evolve with pressure. Concerning the cracking behavior, the fracture of silica glass is assumed to be completely brittle and is thought to be initiated under tensile stress in mode I. All these specificities will be taken into account to model the silica glass mechanical behavior using the developed DEM-constrained natural element method (CNEM) coupling approach. This is the subject of Chapter 5.

5

Modeling of the Silica Glass Mechanical Behavior

5.1. Introduction

Silica glass presents a very complex mechanical behavior compared to classic materials. Under hydrostatic pressure, this material behaves in a perfectly elastic manner up to the densification pressure P_d. The elastic behavior of silica glass can be divided into two parts: nonlinear elastic behavior until a certain limit denoted by P_{nle} and linear elastic behavior beyond P_{nle}. After P_d, silica glass begins to exhibit signs of permanent deformation called densification until the saturation pressure P_s. Beyond P_s, this glass saturates and returns purely elastic. Furthermore, some of its mechanical properties such as Young's modulus, Poisson's ratio and density show an anomalous dependence on the pressure. Although the response of silica glass under hydrostatic pressure is now well known, the shearing effects on this material have been a contentious issue since the first work on the subject [BRI 53]. To address this issue, some researchers have used the numerical approach to investigate the mechanical behavior of this material. Consequently, several numerical models adapted for continuum methods (CMs) have been proposed [RIO 96, MAN 96, BEN 95, HOL 93, JOH 94]. Among them, we can cite the Johnson–Holmquist damage material model, also referred in the literature to as Johnson–Holmquist 2 (JH-2) [HOL 93, JOH 94]. This model is the most used to describe the mechanical behavior of glasses and ceramics with CMs. Nevertheless, as shown in the literature [MIC 07], this model does not allow an accurate modeling of the silica glass mechanical behavior. Furthermore, it is based on continuum mechanics and can only be used for CMs. Despite the current state of the numerical studies on silica glass, modeling the mechanical behavior of this material presents so far a central issue for several scientists.

To properly simulate the laser shock processing (LSP) of this material, the present chapter aims to develop an accurate silica glass model adapted for the proposed DEM-CNEM approach. In this approach, only the discrete element method (DEM) is concerned by the complex mechanical behavior of silica glass. Indeed, this method would be applied in the high-loaded regions, where complex phenomena can occur. The new numerical model of silica glass is then proposed for DEM. On the contrary, CNEM would be applied in the regions far from complex phenomena. Therefore, it would be sufficient to apply a simple elastic model to correctly describe the silica glass response in these constrained natural element method (CNEM) regions. As shown in Chapter 4, different values of the silica glass mechanical parameters (P_{nle}, P_d, P_s, etc.) can be found in the literature. Those retained to develop the new DEM model of silica glass are detailed in section 4.3.3. Using these parameters, a beam-based model adapted for DEM is developed. This model involves several microscopic parameters which must be determined by calibration. To simplify their calibration, the influence of each of these parameters on the resulting macroscopic mechanical behavior is studied. Based on this study, several recommendations simplifying the calibration of these parameters are obtained. The validation of this model is first performed in quasi-statics by simulation of Vickers microindentation, and then in dynamics by simulation of high-velocity impacts of silica glass plates. Concerning the fracture in the DEM regions, the virial-stress-based model developed in the first book of this set [JEB 15] is retained to model the silica glass cracking behavior in these (DEM) regions.

5.2. Mechanical behavior modeling

Using the DEM variation developed in [JEB 15], a continuous domain is discretized into a set of rigid spherical particles connected with cylindrical cohesive beams. The cohesive beams are used to ensure the cohesion of the discretized domain and to model its mechanical behavior. To this end, four microscopic parameters are used to describe the geometrical and elastic mechanical properties of these beams: length l_μ, radius ratio $\tilde{r}_\mu$ (defined as the ratio between the beam radius and the mean radius of the discrete elements), Young's modulus E_μ and Poisson's ration ν_μ. Therefore, one natural approach to model the complex silica glass mechanical behavior by DEM is to enrich the microscopic (elementary) mechanical behavior of these connectors so as to ensure the expected macroscopic (structure) mechanical behavior. The present section details how this idea is put into practice.

5.2.1. *Modeling assumption*

Almost all the works on the densification of silica glass agree that this phenomenon can take place under hydrostatic pressure. However, the role of shearing effects on this phenomenon is contentious and even controversial. Several researchers [COH 61, COH 62, DAC 59] state that the shear stresses have only kinetic effects on the densification and do not change its final level (they only change the rate at which the silica glass responds to pressure), whereas others [ROU 10, CHR 62, MAC 63, WEI 62] hold the opposite view. In this work, it is assumed that the shearing effects on the densification are not significant and are dominated by the hydrostatic pressure effects. Therefore, densification of silica glass is supposed to occur only under hydrostatic pressure. There are several physical and mechanical reasons that support this assumption:

– according to the materials' structure, two main mechanisms can explain the permanent deformation induced by shear stresses. In the case of crystalline solids, plasticity is due to motion of dislocations (crystallographic defects within the regular crystalline structure). As a typical amorphous material, silica glass lacks long-range order (or translational periodicity). Therefore, it is very difficult for dislocations to develop and move within its irregular structure [KAN 98]. In the case of brittle materials, such as rock and concrete, permanent deformation is predominantly caused by nucleation of microdefects. This generally weakens the considered material and degrades its mechanical properties [LUB 08]. This is not the case for silica glass. Permanent deformation of silica glass enhances the mechanical properties [KER 08, JI 07];

– in isotropic materials, the permanent deformation due to shear stresses is volume-conservative. Therefore, even if shear stresses can cause silica glass plasticity, this does not affect the densification level (permanent volume change at a given pressure). For an infinitesimal strain, the volume change can be expressed as:

$$\frac{dV}{V_0} = J - 1 \cong \mathrm{trace}(\varepsilon) \quad [5.1]$$

where dV is an infinitesimal volume variation, V_0 is the initial volume, J is the determinant of the Jacobian matrix [WU 05] and ε is the strain tensor. If only shear stresses are applied, $\mathrm{trace}(\varepsilon) = 0$, and then $dV = 0$.

Based on the assumption that the shearing stresses have no significant effects on the densification of silica glass, a densification model adapted for DEM is developed in this chapter. As will be seen later, comparison between the numerical and experimental densification results proves the validity of this assumption.

5.2.2. *Cohesive beam model*

Using the above modeling assumption, it will be detailed hereafter how the silica glass mechanical behavior is modeled. The key idea of the proposed model is to enrich the mechanical behavior of the cohesive beams in such a way as to ensure the expected macroscopic behavior. This requires linking the microscopic (cohesive beam) and macroscopic (structure) scales. More precisely, it is necessary to find a relationship between the stress loading in the cohesive beams and the resulting macroscopic pressure. To this end, another simplifying assumption is used. This assumption consists of considering that the hydrostatic pressure on a discrete element i is only a function of the normal stresses in the cohesive beams connected to this element:

$$P^i_{hyd} = \sum_{j=1}^{n_N} \sigma^n_{\mu_{ij}} \qquad [5.2]$$

where n_N is the total number of the cohesive beams connected to the discrete element i and $\sigma^n_{\mu_{ij}}$ is the normal stress in the cohesive beam between the discrete elements i and j. Note that the normal stress in a cohesive beam can easily be determined using the Euler–Bernoulli beam theory.

5.2.2.1. *Modeling of nonlinear elasticity*

For relatively low pressures (under approximately $P_{nle} = 3\,GPa$), silica glass shows a nonlinear elastic behavior due to dependence of the bulk modulus K and the shear modulus G on the pressure. Figure 4.13 gives the pressure dependence of these moduli. Based on this figure and using [5.3], relationships between pressure and macroscopic elastic properties (E and ν) can be determined (Figure 5.1):

$$E = \frac{9KG}{3K+G} \quad \text{and} \quad \nu = \frac{3K-2G}{2(3K+G)} \qquad [5.3]$$

To model the nonlinear elasticity of silica glass, the variations of the macroscopic mechanical properties need to be translated into variations of the microscopic (cohesive beam) properties. In other words, it is necessary to find relationships between the macroscopic and microscopic mechanical properties.

As shown in the first book of this set [JEB 15], the parametric study of the cohesive beam parameters yielded the following headings:

– the microscopic Poisson's ratio of the cohesive beams ν_μ has no influence on the macroscopic elastic behavior;

– the macroscopic Poisson's ratio ν depends only on the beam radius ratio $\tilde{r}_\mu$. The function relating these two parameters may be approximated by a second-order polynomial;

– the macroscopic Young's modulus E is a function of both $\tilde{r}_\mu$ and the microscopic (beam) Young's modulus E_μ. For a given value of $\tilde{r}_\mu$, E depends linearly on E_μ, whereas it is a second-order function of $\tilde{r}_\mu$ for a constant E_μ.

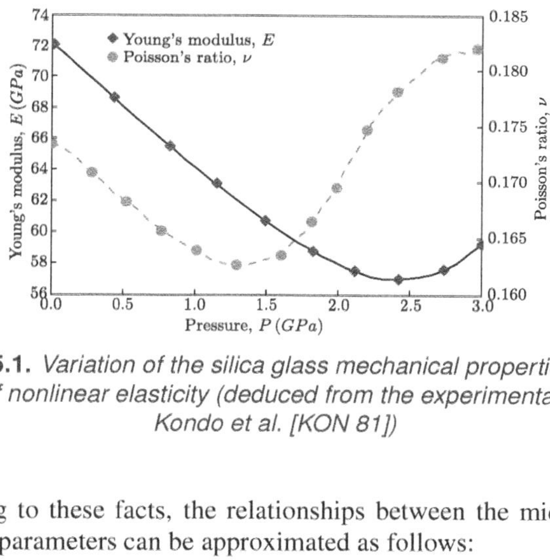

Figure 5.1. *Variation of the silica glass mechanical properties in the region of nonlinear elasticity (deduced from the experimental work of Kondo et al. [KON 81])*

According to these facts, the relationships between the microscopic and macroscopic parameters can be approximated as follows:

$$\tilde{r}_\mu = a_0 + a_1\nu + a_2\nu^2 \qquad [5.4]$$

$$\begin{cases} E_\mu &= \dfrac{E - b_0(\tilde{r}_\mu)}{b_1(\tilde{r}_\mu)} \\ b_0(\tilde{r}_\mu) &= b_{00} + b_{01}\tilde{r}_\mu + b_{02}\tilde{r}_\mu^2 \\ b_1(\tilde{r}_\mu) &= b_{10} + b_{11}\tilde{r}_\mu + b_{12}\tilde{r}_\mu^2 \end{cases} \qquad [5.5]$$

Parameters a_i and b_{ij} of equations [5.4] and [5.5] can be determined numerically using the least squares (LSs) technique. Bearing in mind the simplifying assumption [5.2], the variation of the macroscopic properties with pressure can be translated, at the beam level, into variation of the microscopic properties with normal stress in the beams. The region of nonlinear elasticity can, therefore, be expressed in terms of normal stress in the beams as:

$$\{\sigma_\mu^n | \sigma_\mu^n < \sigma_\mu^{nle}\} \qquad [5.6]$$

where σ_μ^{nle} is the normal stress threshold beyond which the elastic behavior becomes linear. This parameter has to be determined by calibration. Using equations [5.4] and [5.5], the variation of the microscopic properties E_μ and $\tilde{r}_\mu$ in this region [5.6] can be determined. This allows the nonlinear elastic behavior to be taken into account in the silica glass numerical model. Using dimensionless variables $\tilde{\sigma}_\mu^n = \sigma_\mu^n / \sigma_\mu^{nle}$ and $\tilde{P} = P/P_{nle}$, the dimensionless normal stress $\tilde{\sigma}_\mu^n$ is computed for each beam. If $\tilde{\sigma}_\mu^n < 1$, the microscopic properties of the associated beam must be updated. First, the current macroscopic properties associated with $\tilde{P} = \tilde{\sigma}_\mu^n$ must be determined, using Figure 5.1. Then, the corresponding microscopic properties can be evaluated, using [5.4] and [5.5]. These properties can finally be used to update the mechanical behavior of the associated cohesive beam.

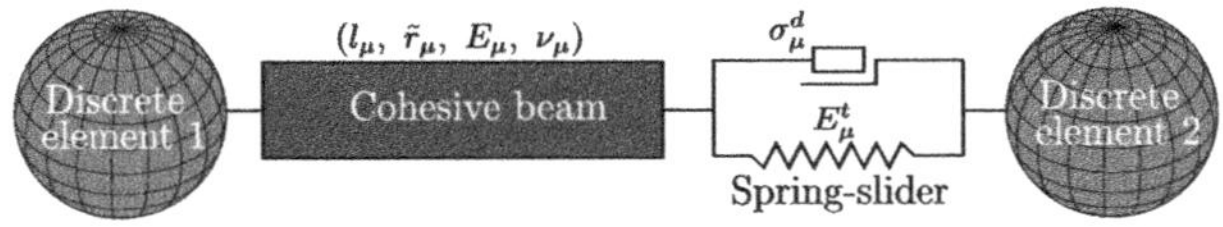

Figure 5.2. *Rheological model of the cohesive bonds used to model densification of silica glass*

5.2.2.2. *Modeling of densification*

To model the densification of silica glass, a "spring-slider" system is connected in series with each cohesive beam linking two discrete elements (Figure 5.2). For each bond in compression, when the compressive normal stress σ_μ^c reaches the microscopic densification pressure σ_μ^d, the slider activates and the densification mechanism takes place. The densification level beyond σ_μ^d is controlled by the microscopic tangential modulus E_μ^t, which is adjustable (Figure 5.3). Finally, to model the saturation stage, the slider stops slipping above a certain limit controlled by the maximum microscopic permanent deformation of the bonds $\varepsilon_\mu^{p\,max}$.

During densification, the macroscopic mechanical properties of silica glass (Young's modulus E and Poisson's ratio ν) vary with pressure (Figure 4.14). For the sake of simplicity, the E and ν variations are approximated in the present work by linear piecewise functions (Figure 5.4).

Beyond P_{nle}, the macroscopic mechanical properties increase linearly with pressure between P_d and P_s (densification and saturation pressure, respectively), and are constant otherwise. To introduce such variations in the densification model, the same technique as detailed in section 5.2.2.1 is applied. Using equations [5.4] and [5.5], the associated microscopic properties variations with normal stress in the beams can be determined and then taken into account in the densification model. Figure 5.5 presents the cohesive bond behavior taking into account the variation of the microscopic mechanical properties with densification. The present densification model involves three microscopic parameters σ_μ^d , E_μ^t and $\varepsilon_\mu^{p\,max}$. These parameters have to be determined by calibration. To better analyze the densification model in the absence of complex dynamic effects, these parameters will first be studied and calibrated in quasi-statics.

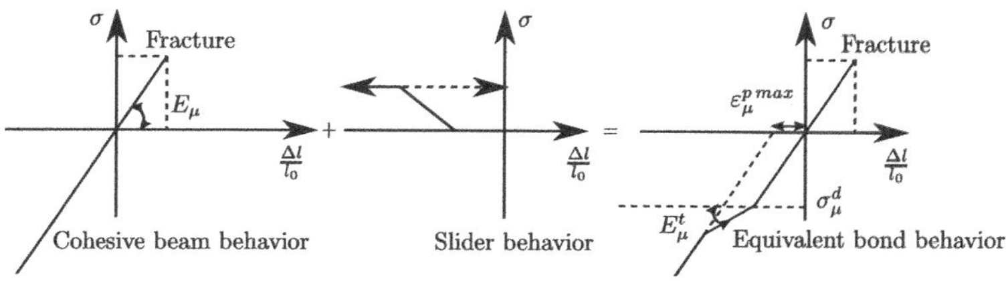

Figure 5.3. *Equivalent cohesive bond behavior*

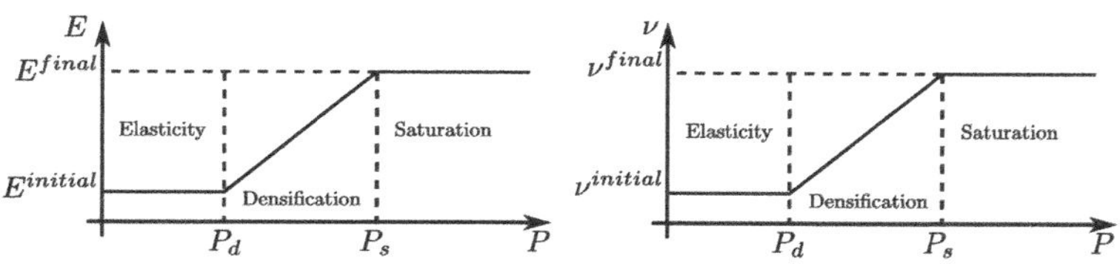

Figure 5.4. *Linear piecewise (C^0) approximation of the variations of the silica glass mechanical properties with compressive pressure*

5.2.3. *Quasi-static calibration and validation*

5.2.3.1. *Calibration*

In the case of quasi-static studies, the nonlinear elastic behavior is not of major importance. Such studies generally focus on the initial and final state of the studied material, e.g. permanent deformation. Therefore, the calibration of the associated microscopic parameter σ_μ^{nle} will be performed in the next section. The present section only details the calibration process of the densification parameters. Three microscopic parameters are involved in the densification model: the microscopic densification pressure σ_{μ}^{d}, the microscopic tangential modulus E_{μ}^{t} and the maximum permanent deformation of the beam bonds $\varepsilon_{\mu}^{p\,max}$. Calibration of such parameters aims to find their appropriate values that ensure the expected macroscopic behavior of silica glass. This behavior is characterized by three macroscopic parameters section 4.3.1: the densification pressure P_d, the saturation pressure P_s and the volume change ratio $\Delta V/V_0$ (or density change ratio $\Delta\rho/\rho_0$). To simplify the calibration process, the influence of each microscopic parameter on the macroscopic ones is studied. A hydrostatic compression of a sphere having $100\,mm$ diameter is used in this study.

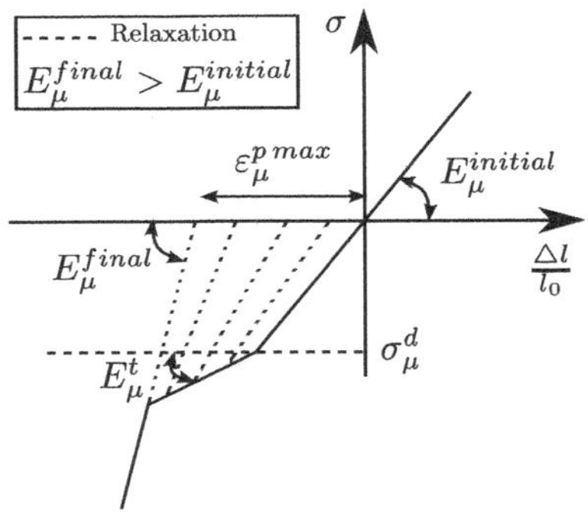

Figure 5.5. *Behavior of equivalent cohesive bond, taking into account the variation of the mechanical properties with densification*

One major problem of the discrete methods is the dependence of results on the number of the discrete elements (n_D) used to discretize the numerical sample. Therefore, the influence of this parameter on the densification response is first studied. The sphere model is discretized using various numbers of discrete elements n_D (from 100 to 20 000). The microscopic densification parameters were fixed as follows: the microscopic densification pressure $\sigma_{\mu}^{d} = 5\,GPa$, the microscopic tangential modulus $E_{\mu}^{t} = 0\,GPa$ and the maximum permanent deformation of the beam bonds $\varepsilon_{\mu}^{p\,max} = 0.05$. Figure 5.6 presents the results of this study. For a small number of discrete elements, the densification response greatly fluctuates with this parameter (Figure 5.6(a)). These fluctuations can be reduced by increasing n_D. Beyond $n_D = 10\,000$, the densification response becomes very weakly affected by the discrete element number (Figure 5.6(b)). This value of n_D is in agreement with the results of the elastic behavior obtained [AND 13]. It is shown in this book that beyond $n_D = 10\,000$, the homogeneity and isotropy of the discrete domain (DEM domain) are ensured and the elastic behavior becomes nearly independent of n_D.

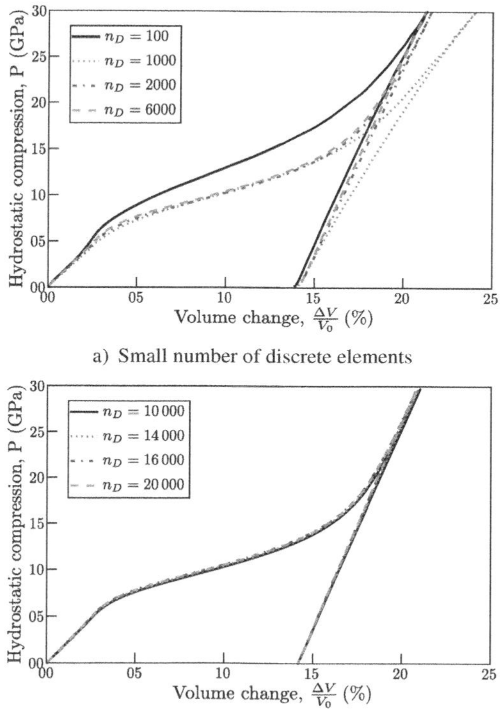

a) Small number of discrete elements

b) Great number of discrete elements

Figure 5.6. *Sensitivity of the macroscopic densification behavior to the number of discrete elements*

In the remainder of this study, $n_D = 10\,000$ discrete elements will be used to discretize the numerical sample. Figure 5.7(a) shows that the macroscopic densification pressure P_d depends linearly on the microscopic one σ_μ^d. In addition, this parameter (P_d) is practically independent of the other microscopic properties, i.e. E_μ^t and $\varepsilon_\mu^{p\,max}$, as observed in Figures 5.7(b) and 5.7(c). Figure 5.8 shows the influence of the various microscopic parameters on the macroscopic saturation pressure P_s. This parameter (P_s) depends linearly on the microscopic densification pressure σ_μ^d, it also slightly depends on the microscopic tangential modulus E_μ^t (around 12% P_s variation for E_μ^t up to $100\,GPa$). However, it is effectively independent of the maximum permanent deformation of the beam bonds $\varepsilon_\mu^{p\,max}$ (Figure 5.8(c)), only 5% P_s

variation for $\varepsilon_{\mu}^{p\,max}$ up to 0.12. Concerning the third macroscopic parameter $\Delta V/V_0$, Figure 5.9 shows that this parameter depends only on E_{μ}^{t} and $\varepsilon_{\mu}^{p\,max}$, and it is effectively independent of σ_{μ}^{d}. After studying the influence of the microscopic parameters on the macroscopic densification behavior, calibration becomes fairly easy. Because the macroscopic densification pressure P_d depends only on the microscopic one σ_{μ}^{d}, this parameter must be calibrated first. Subsequently, the microscopic tangent modulus E_{μ}^{t} can be calibrated to ensure the expected value of the macroscopic saturation pressure. Finally, the maximum permanent deformation of the beam bonds $\varepsilon_{\mu}^{p\,max}$ can be set to find the appropriate global volume change ($\Delta V/V_0$). Table 5.1 gives the calibrated microscopic densification parameters that must be used to ensure the correct macroscopic densification behavior of silica glass.

5.2.3.2. *Validation*

Using the calibrated microscopic densification parameters (Table 5.1), the effectiveness of the proposed densification model will be investigated hereafter at both macroscopic and microscopic scales.

$\sigma_{\mu}^{d}\,(GPa)$	$E_{\mu}^{t}\,(GPa)$	$\varepsilon_{\mu}^{p\,max}$
7.13	98.82	0.067

Table 5.1. *Static microscopic parameters of silica glass mechanical behavior (obtained by calibration)*

5.2.3.2.1. Macroscopic validation

To validate the calibrated densification model at the macroscopic scale, it is applied to simulate several hydrostatic compression tests. The same geometric sample as for the calibration study is used in these tests. This sample is subjected to various hydrostatic pressures. The associated permanent deformations are measured and compared to Ji's [JI 07] experimental results obtained using thick samples (Figure 5.10). As shown in Figure 5.10, the numerical results are in agreement with the experimental results. Therefore, it can be argued that the proposed densification model allows for good estimate of the global densification level of silica glass. However, this may not automatically imply the ability of such a model to reproduce correctly the densification mechanisms at the microscopic scale. To investigate this last point, the densification model is applied to simulate Vickers microindentation tests on silica glass.

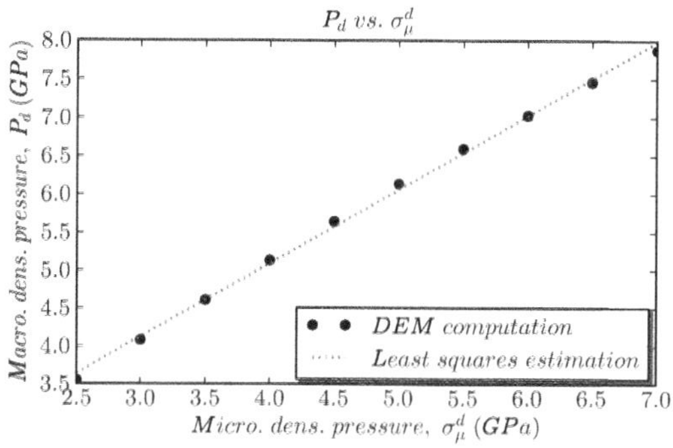

a) $E^t_\mu = 0\,GPa$, $\varepsilon^{p\,max}_\mu = 0.05$

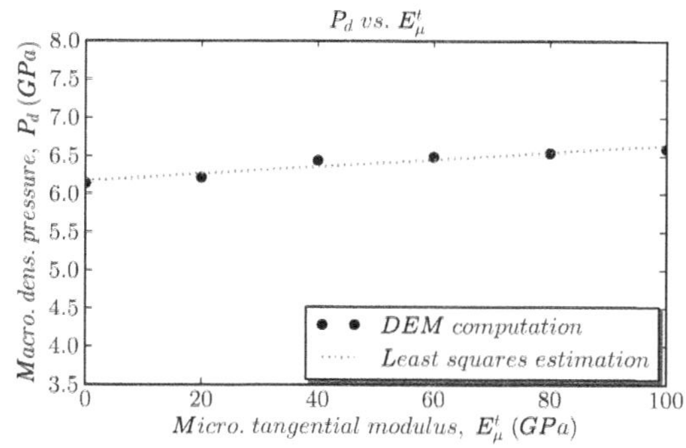

b) $\sigma^d_\mu = 5\,GPa$, $\varepsilon^{p\,max}_\mu = 0.05$

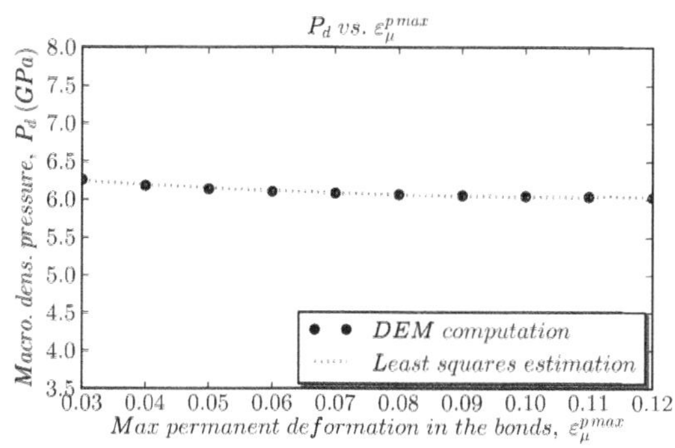

c) $\sigma^d_\mu = 5\,GPa$, $E^t_\mu = 0\,GPa$

Figure 5.7. *Variation of macroscopic densification pressure (P_d) with the microscopic densification parameters: microscopic densification pressure (σ^d_μ), microscopic tangential modulus (E^t_μ) and maximum permanent deformation of the beam bonds ($\varepsilon^{p\,max}_\mu$)*

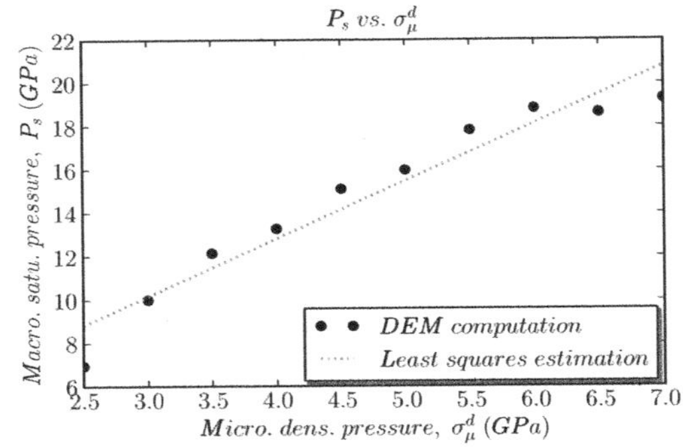

a) $E_\mu^t = 0\,GPa,\ \varepsilon_\mu^{p\,max} = 0.05$

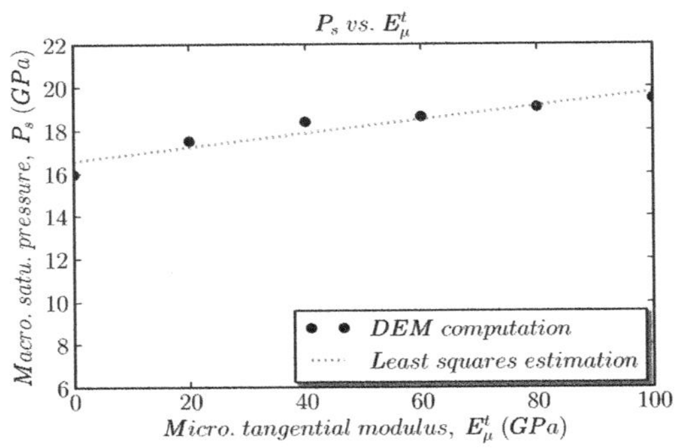

b) $\sigma_\mu^d = 5\,GPa,\ \varepsilon_\mu^{p\,max} = 0.05$

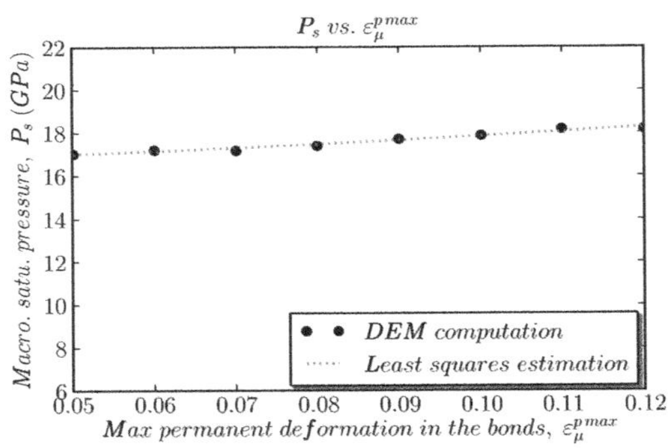

c) $\sigma_\mu^d = 5\,GPa,\ E_\mu^t = 0\,GPa$

Figure 5.8. *Variation of macroscopic saturation pressure (P_s) with the microscopic densification parameters: microscopic densification pressure (σ_μ^d), microscopic tangential modulus (E_μ^t) and maximum permanent deformation of the beam bonds ($\varepsilon_\mu^{p\,max}$)*

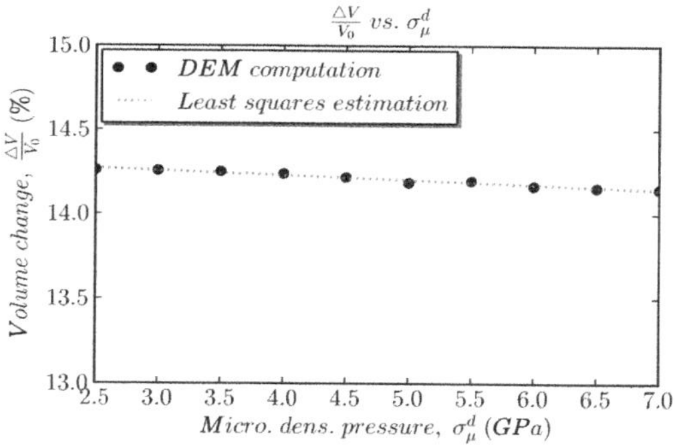

a) $E_\mu^t = 0\,GPa$, $\varepsilon_\mu^{p\,max} = 0.05$

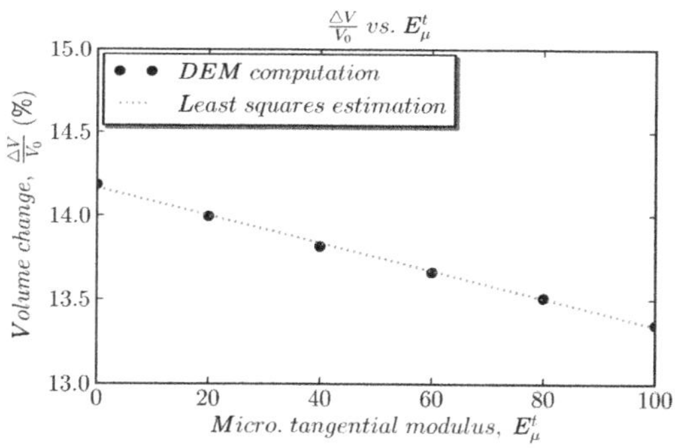

b) $\sigma_\mu^d = 5\,GPa$, $\varepsilon_\mu^{p\,max} = 0.05$

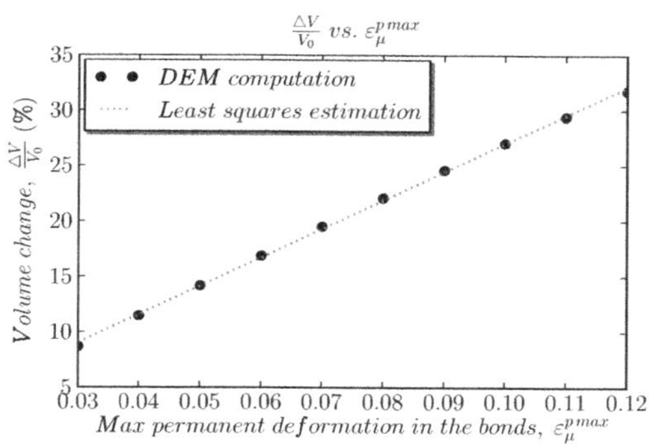

c) $\sigma_\mu^d = 5\,GPa$, $E_\mu^t = 0\,GPa$

Figure 5.9. *Variation of volume change ($\frac{\Delta V}{V_0}$) with the microscopic densification parameters: microscopic densification pressure (σ_μ^d), microscopic tangential modulus (E_μ^t) and maximum permanent deformation of the beam bonds ($\varepsilon_\mu^{p\,max}$)*

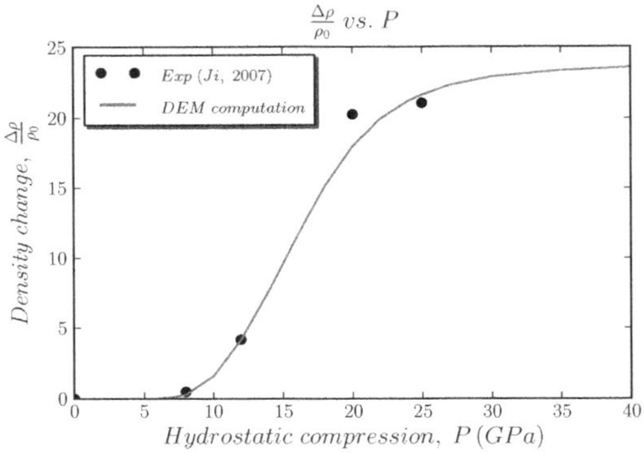

Figure 5.10. *Permanent density change with pressure (ex-situ measures): comparison between experimental [JI 07] and numerical results.*

5.2.3.2.2. Microscopic validation

Silica glass undergoes permanent densification under high compressive pressure. Comparable stresses can be reached beneath a sharp indenter (e.g. Vickers indenter) during an indentation test. To verify the validity of the developed densification model at the microscopic scale, Vickers microindentations are simulated. Low indentation forces are used to study the densification behavior in the absence of fracture: $F_{indent} = 0.1\,N$ and $F_{indent} = 0.5\,N$. A hemispherical DEM domain of $12\,\mu m$ radius is used in the present simulations (Figure 5.11). This domain, whose the spherical part (dome) is fixed, is discretized using $n_D = 100\,000$ discrete elements (i.e. the average radius of the discrete elements is around $0.2\,\mu m$) to capture the microscopic mechanisms of the densification. The top surface of this domain is indented with a square-based pyramidal indenter with included face angles of 136° (Vickers indenter) and $2\,\mu m$ in height (Figure 5.11). This indenter undergoes a moderate displacement in the z direction until the reacting force reaches the desired indentation force. Table 5.2 presents a comparison between the experimental [JI 07, YOS 05] and the present numerical indentation results for low indentation forces ($F_{indent} = 0.1\,N$ and $F_{indent} = 0.5\,N$).

For $F_{indent} = 0.1\,N$, the DEM simulation yielded a relatively good estimate of the indentation print volume V^- and the volume of the piled-up material around the indenter V^+. The densified volume V^d which is a function of V^- and V^+ compares favorably with the experimental results [JI 07, YOS 05] (in comparison to previous numerical studies [JI 07]).

Furthermore, the diagonal length of the indentation print (L) is well estimated numerically. It should be noted that this parameter is an important indicator of the material hardness. However, the indentation depth D is undervalued, compared to the experimental results [JI 07, YOS 05]. This can be explained by the fact that in the experimental tests, microcracks can initiate under the indenter tip (zone of high stress concentration), causing measurement errors in D. The experimental value of D can include not only the depth of the indentation print but also the depth induced by the cracks.

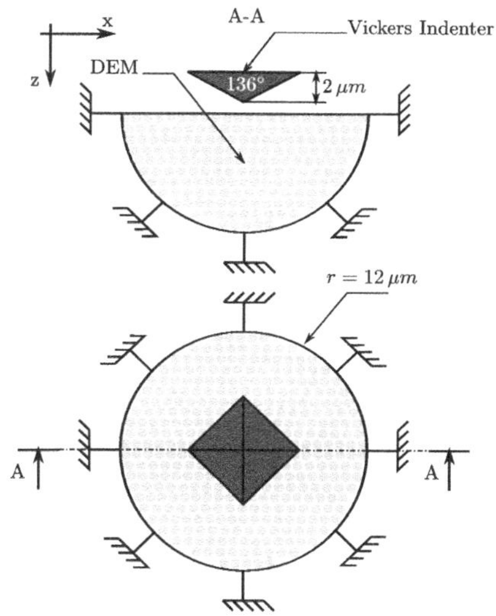

Figure 5.11. *DEM indentation model*

For $F_{indent} = 0.5\,N$, the densification state becomes more significant and a much greater densification print is obtained (Figure 5.13). Compared to the experimental results (Table 5.2), the DEM numerical simulation provides a relatively good estimate of the volume of the piled-up material around the indenter V^+ and the diagonal length L of the indentation print. Concerning D, the analysis of the $F_{indent} = 0.1\,N$ case remains valid. However, the gap between the numerical and experimental values of V^- is more significant than for $F_{indent} = 0.1\,N$. This may be due to side-effects. In the case of $F_{indent} = 0.5\,N$, the densified zone is relatively important compared to the global geometric model. Consequently, the effects of the overstated conditions applied to the dome (the spherical part of the geometric model) become significant. This can increase the region where the densification pressure P_d can be reached, hence the overstatement of V^-.

One way to reduce the side-effects is to use a larger DEM domain. However, such a solution can increase considerably the computational time, because a very fine discretization must be applied to capture the microscopic densification effects. Another solution less costly is to apply the developed DEM-CNEM coupling approach. Using this approach, DEM is applied in the indentation region where the densification pressure (P_d) can be reached, whereas CNEM is applied in the remainder of the modeled domain. Figures 5.12 present the coupling model which is a box of $30\,\mu m$ sides length. The DEM subdomain is the same as in the previous study (fully DEM computation). Beyond the indentation region, 10 000 CNEM nodes are used to discretize the rest of the studied domain (Figure 5.12). Based on the coupling parameters study section 3.3, a $2\,\mu m$–thick bridging region is used between the coupled models to ensure a correct wave propagation between them. Since the CNEM domain is far from the indentation region, only an elastic behavior is applied in this region.

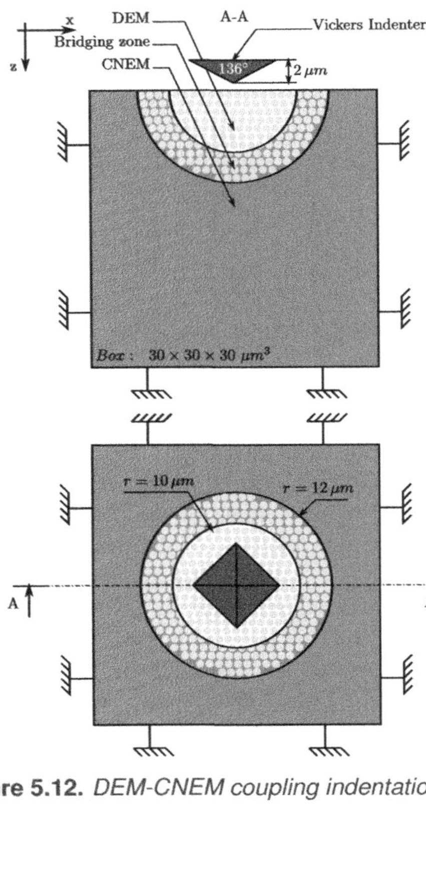

Figure 5.12. *DEM-CNEM coupling indentation model*

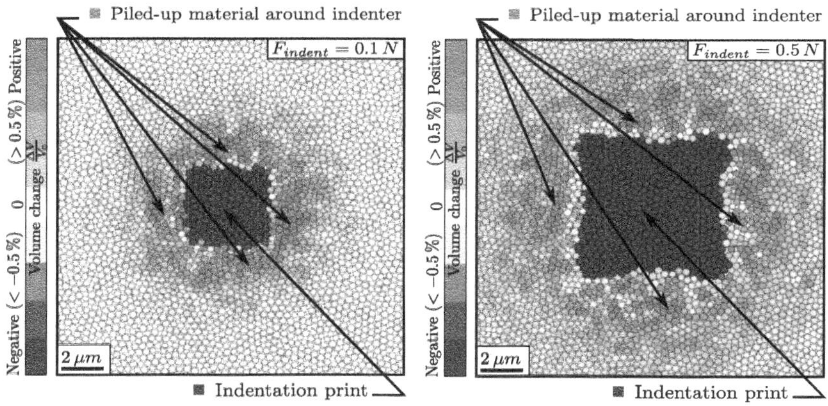

Figure 5.13. *Top view of the indentation print and the piled-up material around the indenter. a): $F_{indent} = 0.1\,N$; b): indentation force $F_{indent} = 0.5\,N$. For a color version of the figure, see www.iste.co.uk/jebahi/discrete2.zip*

As shown in Table 5.2, in the case of $F_{indent} = 0.1\,N$, the DEM and DEM-CNEM coupling results are practically the same. However, in the case of $F_{indent} = 0.5\,N$ where the side-effects become significant, the DEM-CNEM coupling results are slightly better than those obtained with fully DEM computation. Nevertheless, these results are still relatively overestimated compared to the experimental ones. As shown in Figure 5.14, application of the DEM-CNEM coupling improved the shape of the densified zone. It can be observed that the densified zone obtained from the coupling method (Figure 5.14(c)) is closer to that obtained experimentally by Ji [JI 07] (Figure 5.14(a)) than the densified zone obtained from the DEM computation (Figure 5.14(b)). Indeed, it is less deep and broader in the former (coupling result) than in the latter (DEM result).

Overall, the densification results obtained using the developed silica glass model compare favorably with the experimental results of Ji [JI 07]. This model gives an acceptable quantitative estimate of silica glass densification at both the microscopic and macroscopic scales.

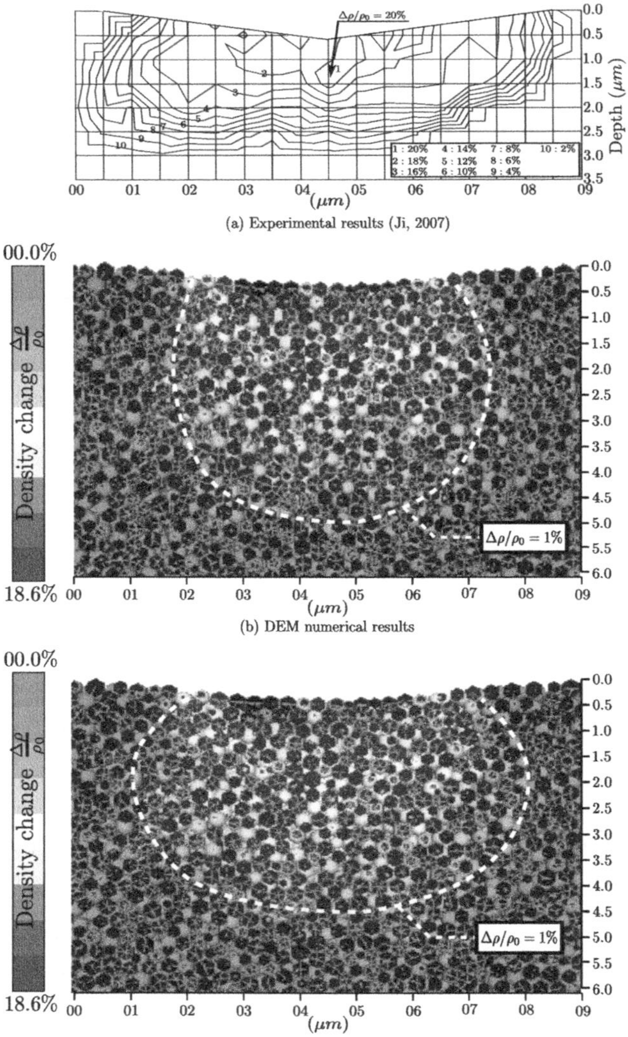

Figure 5.14. *Cut-away view of the indented region;* $F_{indent} = 0.5\,N$*; (a) experimental results from [JI 07], (b) DEM numerical results and; (c) DEM-CNEM coupling numerical result. For a color version of the figure, see www.iste.co.uk/jebahi/discrete2.zip*

F_{indent}		V^- (μm^3)	V^+ (μm^3)	V^d (μm^3)	D (μm)	L (μm)
	Num: DEM	1.304	0.136	1.168	0.159	4.758
$0.1\,N$	Num: Coupling	1.302	0.131	1.171	0.159	4.758
	Num [JI 07]	$--$	$--$	0.058	$--$	$--$
	Exp [JI 07]	0.727	0.108	0.605	0.377	4.707
	Exp [YOS 05]	1.034	0.018	0.952	0.421	5.157
	Num: DEM	14.534	1.174	13.36	0.397	9.318
$0.5\,N$	Num: Coupling	13.628	0.920	12.708	0.409	9.957
	Num [JI 07]	$--$	$--$	0.61	$--$	$--$
	Exp [JI 07]	9.971	1.138	8.805	0.908	10.371
/	Exp error [JI 07]	$\pm 4\%$	$\pm 7\%$	$\pm 10\%$	$\pm 6\%$	$\pm 2\%$

$V^-$and V^+ are, respectively, the volume of the indentation print and the volume of the piled-up material around the indent, V^d is the densified volume, D is the indentation depth and L is the diagonal length of the indentation print.

Table 5.2. *Vickers indentation results: low indentation forces*

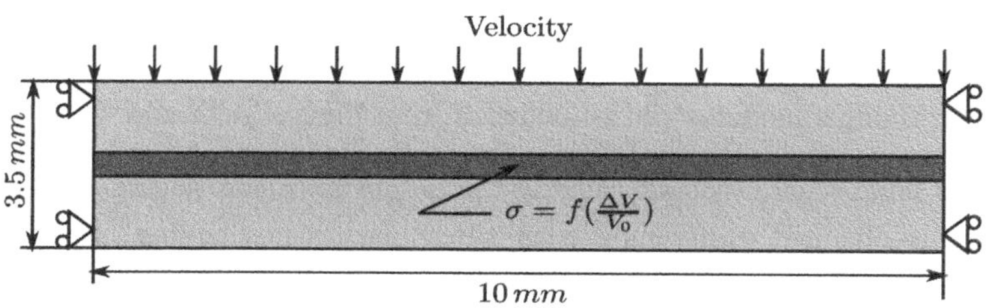

Figure 5.15. *Geometric model for the dynamic calibration of the microscopic nonlinear elastic and densification parameters*

5.2.4. *Dynamic calibration and validation*

In the previous section, the silica glass model is studied in quasi-statics to better analyze its effectiveness, in terms of predicting the densification mechanisms at both the macroscopic and microscopic scales, in the absence of complex dynamic effects. The present section aims to study this model in dynamics which is the scope of this work.

5.2.4.1. *Calibration*

To calibrate the silica glass numerical model in dynamics, a plate of $10 \times 10\,mm^2$ square base and $3.5\,mm$ thickness is used. The top surface of the plate is subjected to high-velocity loadings up to $4000\,ms^{-1}$ (Figure 5.15). To ensure a state of uniaxial strain, symmetric boundary conditions are imposed on the lateral sides. The plate is discretized using $n_D = 50\,000$ discrete elements.

To model the nonlinear elastic behavior of silica glass in the region of low pressures, only one microscopic parameter is used: the normal stress threshold σ_{μ}^{nle} from which the silica glass behavior returns linear elastic. This parameter is easy to obtain and can be calibrated first, since it is independent of the densification parameters. To this end, a series of simulations using different values of σ_{μ}^{nle} is performed until obtaining the expected macroscopic P_{nle}. Table 5.3 gives the calibrated value of σ_{μ}^{nle}. Thus, the densification parameters can be calibrated. The calibration of these parameters must be performed with respect to the Hugoniot curve of silica glass (Figure 5.17). This curve characterizes the mechanical behavior of this material in (highly) dynamics. Concerning the variations of the mechanical properties with pressure in the region of densification, only the variation of the Young's modulus with pressure is taken into account in the dynamic studies. No experimental data are found concerning other properties, such as Poisson's ratio. Considering a constant Poisson ratio $\nu = 0.17$, the pressure dependence of the Young's modulus can be deduced from Figures 4.17 and 4.18, which show, respectively, the evolution of the bulk modulus and the longitudinal stress with density change ($\Delta\rho/\rho_0$). It should be noted that the Hugoniot curve is obtained using impacts of silica glass plates having large lateral dimensions compared to the thickness. Therefore, the state of uniaxial stress in the plate center (measuring region) can also be viewed as a state of uniaxial strain. In this case, the longitudinal stress can be considered as a pressure. Figure 5.16(a) presents the longitudinal stress dependence of Young's modulus. This figure can also be regarded as pressure dependence of Young's modulus. For the sake of simplicity, the pressure dependence of Young's modulus is approximated by linear piecewise function (Figure 5.16(b)). E linearly increases between P_d and P_s and remains constant everywhere else.

Table 5.3 gives two proposals of calibrated densification parameters. The first proposal correctly fits the elastic and densification domains of silica glass; however, it overestimates the macroscopic density change $\Delta\rho/\rho_0$ (Figure 5.17). The second proposal corrects this problem, but it loses accuracy at the end of the densification domain as can be seen in Figure 5.17. Since this work aims to simulate the LSP processing on silica using high-power lasers, of which the induced pressures can far exceed P_s, the second proposal is retained for this model.

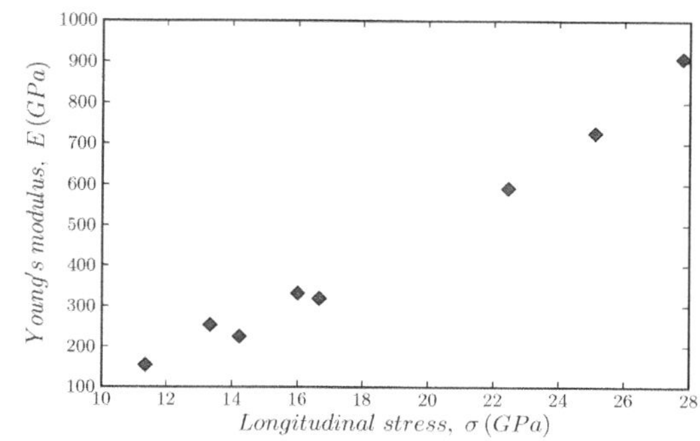

a) Experimental result deduced from [LAL 91]

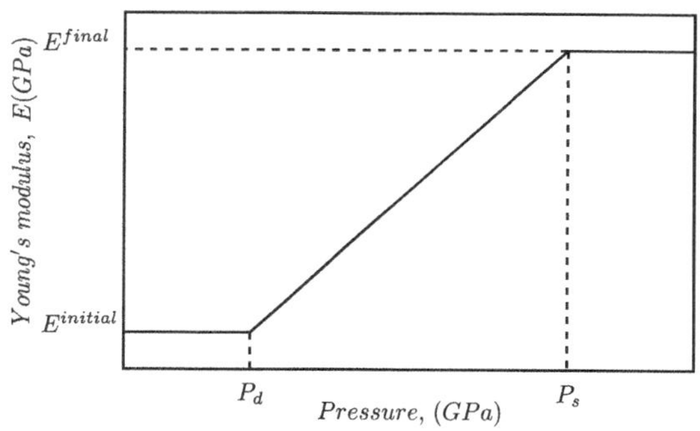

b) Linear piecewise approximation for DEM modeling

Figure 5.16. *Young's modulus variation in dynamics, deduced from the experimental results of Lalle et al. [LAL 91]*

		σ_μ^{nle} (GPa)	σ_μ^{d} (GPa)	E_μ^{t} (GPa)	$\varepsilon_\mu^{p\,max}$
Quasi-statics		$--$	7.13	98.82	0.067
Dynamics	Proposal 1	2.5	12	20	0.34
	Proposal 2	2.5	12	20	0.325

Table 5.3. *The microscopic parameters of silica glass mechanical behavior (obtained by calibration)*

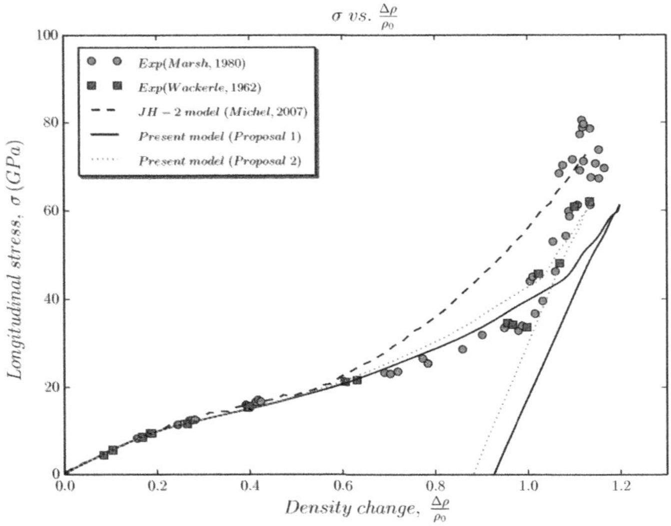

Figure 5.17. *Hugoniot curve of silica glass: comparison between present numerical results and other experimental [MAR 80, WAC 62] and numerical [MIC 07] results from the literature*

5.2.4.2. *Validation*

To validate the calibrated densification model of silica glass in dynamics, the high-velocity impact tests performed by Sugiura *et al.* [SUG 81] of silica glass plates are reproduced numerically in this work. In these tests, both the flayer and target plates are made of silica glass (symmetrical impact). The final particle velocity is expected to be exactly half of the impact velocity, assuming that there is no diminution in particle velocity at any point. Therefore, to reduce the computation time, only a small portion (located at the center) of the target with symmetric lateral boundary conditions is considered in this simulation (Figure 5.19). The top surface (impact surface) is subjected to half of the impact velocity (Figure 5.19). With these considerations, the validation geometric model is quite similar to that used for the dynamic calibration. To reduce the velocity jump at the first time step (for numerical stability requirements), a small initial slope is used to apply the velocity loading (Figure 5.18). In order to compare the numerical results with the experimental results, two measuring points A and B are introduced in the geometric model. These points are, respectively, located at $1\,mm$ and $3\,mm$ distance from the top surface (Figure 5.19).

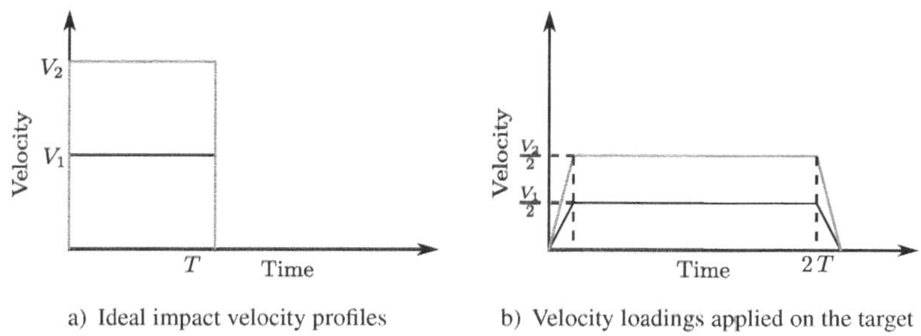

Figure 5.18. *Impact velocity modeling; V_1 and V_2 are the two different impact velocities*

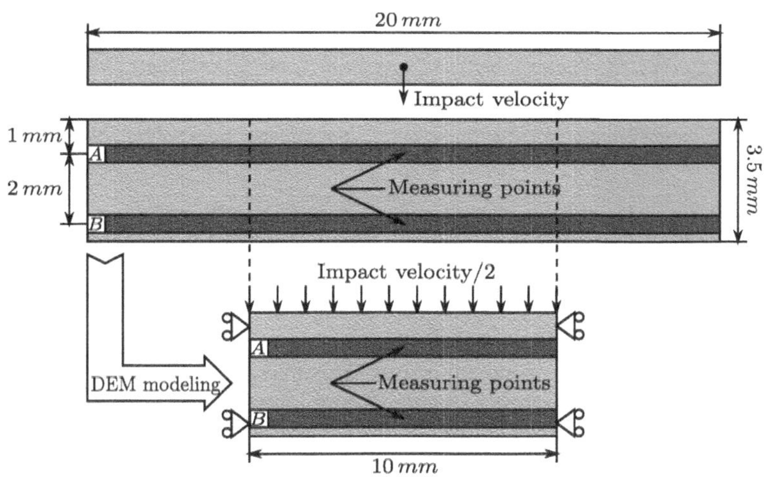

Figure 5.19. *Geometric model of the high-velocity impact tests*

Figure 5.20 presents a comparison between the experimental [SUG 81] and numerical particle velocity profiles in the measuring points A and B (Figure 5.19), for an impact velocity $V_{Impact} = 3310\,ms^{-1}$. This figure provides only a qualitative comparison of the results. The experimental particle velocity profiles are presented as recorded on the oscilloscope, and then are given in terms of voltage versus time. Qualitatively, the numerical results are in good agreement with the experimental results. The different specificities of the silica glass response under high-velocity impacts are properly captured with the proposed silica glass model.

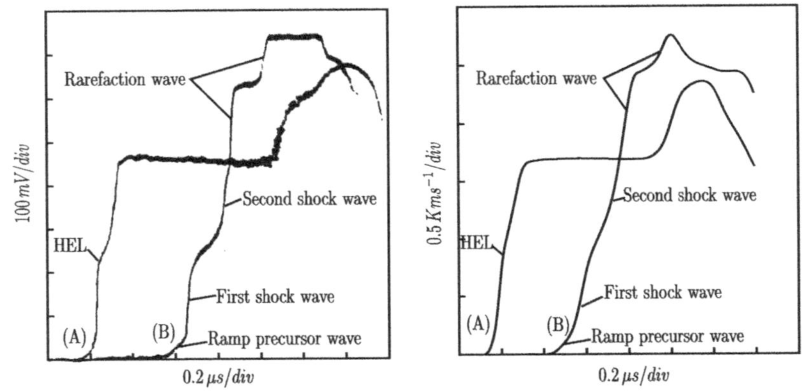

a) Experimental result from [SUG 81]   b) Numerical result

Figure 5.20. *Qualitative comparison between numerical and experimental [SUG 81] particle velocity profiles in the densification region (impact velocity $V_{Impact} = 3310 ms^{-1}$); curves (A) and (B) correspond, respectively, to the front and rear measuring points*

Quantitatively, Figure 5.21 shows a comparison between the numerical and experimental results for different impact velocities: $880\, ms^{-1}$ (elastic region), $2580\, ms^{-1}$ (densification region: near to the Hugoniot elastic limit (HEL)) and $3310\, ms^{-1}$ (densification region: far from the HEL). Of particular note is that the shock fronts (first and second shock front) are not very steep as in the experimental results. This may be due to the discretization fineness (number of discrete elements used to discretize the probem domain). Indeed, the shock front (even physically) is not an exact discontinuity but it spreads out over several layers of discrete elements (atoms). Therefore, a very fine discretization is required to obtain steep shock fronts. In the elastic region ($V_{Impact} = 880\, ms^{-1}$), a relatively good estimate of the maximum particle velocity V_p^{max} is obtained. However, in the densification region, slight difference between the experimental and numerical values of V_p^{max} is observed. The numerical model underestimates V_p^{max}. The discrepancy between the numerical and experimental values of V_p^{max} increases as the impact velocity increases. This can be explained by the fact that the initial slope used to apply the impact velocity V_{Impact} (Figure 5.18(b)) steepens up as the amplitude of V_{Impact} increases. Therefore, additional very high-frequency waves (HFWs) exceeding the cutoff frequency of the DEM discretization can be involved. These HFWs are not supported by the DEM discretization and are numerically filtered, resulting in V_p^{max} undervaluation. Another reason which can explain the V_p^{max} understatement is that the equation of state (Hugoniot) given in Figure 5.17 is satisfied only at larger scale and not at each discrete element. Therefore, some errors can be introduced in the numerical estimate of V_p^{max}, which is computed as the average velocity of the discrete elements belonging to the measuring region. Table 5.4 gives a comparison between the numerical and experimental shock waves velocities (elastic and densification waves) for different V_{Impact}. The numerical model provides a good estimate of the 1st shock front velocity for all V_{Impact}. Concerning the 2nd shock front velocity, the numerical results are somewhat less good. The proposed silica glass model overestimates this velocity, near to the HEL (case of $V_{Impact} = 2580\,ms^{-1}$), and underestimates it, far from this limit (case of $V_{Impact} = 3310\,ms^{-1}$). To explain this, the calibrated Hugoniot curve and that obtained by Sugiura *et al.* [SUG 81] (whose experimental results are used here) are isolated and zoomed in on the region of densification (Figure 5.22). Near to the HEL, the slope of the numerical Hugoniot curve is greater than the slope of the experimental one, whereas the opposite trend is shown far from the HEL. Remember that the velocity of a disturbance in matter is proportionate to $\sqrt{(\partial P/\partial \rho)_s}$. This explains the overstatement of the the 2nd shock front velocity for $V_{Impact} = 2580\,ms^{-1}$ and the understatement of this velocity for $V_{Impact} = 3310\,ms^{-1}$. Overall, the numerical results would be acceptable. The discrepancy between the experimental and numerical results does not exceed 9%, which is the estimation error of the 2nd shock front velocity for $V_{Impact} = 2580\,ms^{-1}$ (Table 5.4).

V_{Impact}		1st shock (elastic) front		2nd shock (plastic) front	
(ms^{-1})		Shock velocity (ms^{-1})	Particle velocity (ms^{-1})	Shock velocity (ms^{-1})	Particle velocity (ms^{-1})
880	Num	4931	434	–	–
	Exp [SUG 81]	4940	440	–	–
2580	Num	4884	802	4230	1275
	Exp [SUG 81]	5030	750	3910	1290
3310	Num	4901	806	4300	1630
	Exp [SUG 81]	4990	800	4550	1655

Table 5.4. *Shock fronts velocities for different impact velocities*

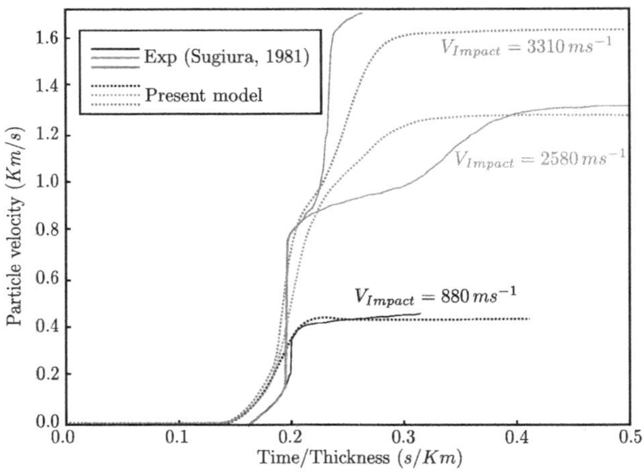

Figure 5.21. *Particle velocity profiles corresponding to impact velocities of $880\,ms^{-1}$, $2580\,ms^{-1}$ and $3310\,ms^{-1}$: comparison between numerical and experimental results obtained by Sugiura et al. [SUG 81]. For a color version of the figure, see www.iste.co.uk/jebahi/discrete2.zip*

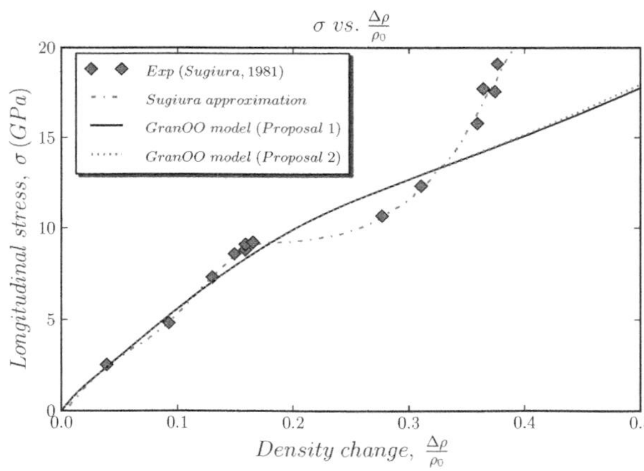

Figure 5.22. *Comparison between numerical and Sugiura's Hugoniot curves in the region of densification (experimental data taken from [SUG 81])*

5.3. Brittle fracture modeling

As a brittle material, silica glass presents a brittle fracture behavior. This type of fracture is thought to be initiated under tensile stress in mode I [GRI 21, WIE 04]. This section details how this fracture behavior is modeled by DEM.

As reported in [JEB 15], DEM can naturally deal with complex cracking behaviors. Consequently, it has become widely used to model complex fracture problems [RIC 07, IOR 08, TAN 09, MAG 98, SHI 09, D'AD 01]. The first book of this set [JEB 15] discussed the ability of the existing fracture models to quantitatively fit the experimental cracking mechanisms [LAW 03]. It demonstrated that these models give acceptable results at the macroscopic scale. However, these cannot reproduce the cracking mechanisms at the microscopic scale. To overcome this limitation, that book [JEB 15] proposed a non-local fracture model based on the computation of an equivalent Cauchy stress tensor $\boldsymbol{\sigma}$ at the discrete elements. This model postulates that a discrete element is released from its neighbors when the associated hydrostatic pressure (computed from $\boldsymbol{\sigma}$) is tensile and exceeds the discrete element fracture strength σ_{hyd_f}. The appropriate value of σ_{hyd_f} must be determined by calibration. Application of this model to simulate various challenging brittle fracture problems yielded very promising results. Therefore, it was retained in this work to model the complex cracking behavior of silica glass. As mentioned in the literature [JEB 13b, ARO 79, GRO 12], this complex behavior presents some particularities due to densification. Generally, classic materials have a tendency to produce cone cracks under a spherical indenter. However, when a sharp indenter (e.g. Vickers indenter) is used, these materials produce median, radial and lateral cracks. On the contrary, silica glass has a strong tendency to form a cone crack, even when indented with a sharp tip. Because of its important densification behavior, a hemispherical densified region is formed beneath the sharp indenter which in turn operates as a spherical indenter on the rest of the problem domain. A cone crack may, therefore, set up which can accompany median, radial and lateral cracks. This particularity is not encountered with other types of glasses and is very difficult to capture numerically. In order to investigate the effectiveness of the retained fracture model, it was applied in an attempt to reproduce this silica glass particularity. To this end, Vickers microindentation is simulated using high indentation force ($30\,N$). The same model presented in Figure 5.11 is used in this simulation. Since the cracking process is expected to be initiated from the surface, $\sigma_f = 50\,MPa$ was used as silica glass fracture strength to calibrate the discrete element fracture threshold σ_{hyd_f}. Calibration of this parameter which was discussed in [JEB 15] yielded $\sigma_{hyd_f} = 64\,MPa$.

Figure 5.23 presents the numerical cracking response of silica glass indented with a Vickers tip at the beginning of fracture (before the cracking becomes unstable). In this figure, only the discrete elements where the fracture criterion is reached are shown. The obtained results are in good agreement with the experimental results [ARO 79, GRO 12]. The radius of the numerical cone crack ($r_{num}^{cone} = 41.14\,\mu m$) is very close to that obtained experimentally by Arora *et al.* [ARO 79] ($r_{exp}^{cone} \approx 40\,\mu m$). This proves the validity of the retained fracture model to simulate the complex silica glass cracking behavior.

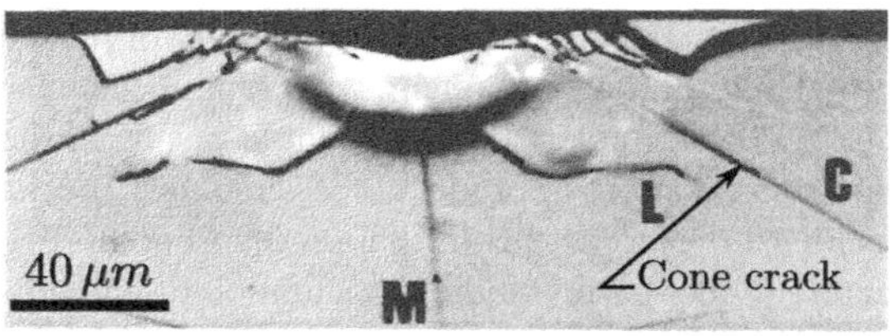

(a) Experimental result (Arora, 1979)

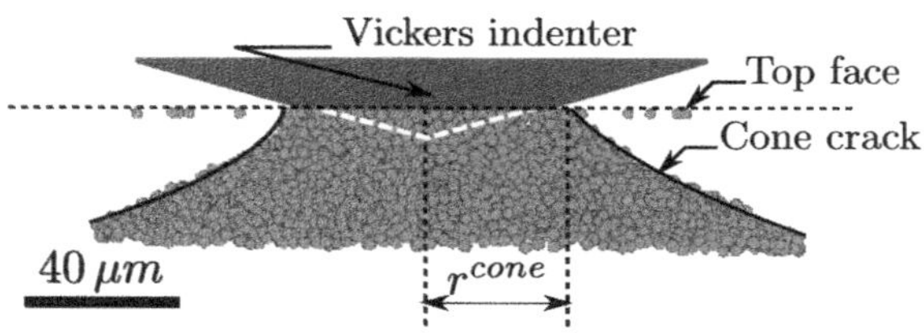

(b) Cone crack: side view

(c) Cone crack: perspective view

Figure 5.23. *Cracking pattern of silica glass indented with a Vickers tip; $F_{indent} = 30\,N$; only the discrete elements where the fracture criterion is reached are shown in (b) and (c)*

5.4. Conclusion

To correctly simulate the LSP of silica glass, the present chapter dealt with the modeling of the complex mechanical behavior of this material using the developed DEM-CNEM coupling approach. In this approach, CNEM would be applied in the regions far from complex LSP phenomena. Therefore, only elastic behavior is considered in these regions. On the contrary, DEM would be applied in the laser-affected zones. To correctly model the silica glass response in these zones, a new silica glass model adapted for DEM was proposed. This model is based on an assumption that the shearing effects (matter of controversy in the literature) are negligible, compared to pressure effects.

To model the silica glass nonlinear elastic behavior in the region of low pressures, empirical relationships were proposed to translate the macroscopic variations of the silica glass mechanical properties into microscopic variations of the cohesive beam parameters. To model the silica glass densification, the original elastic cohesive beams between discrete elements are connected in series with "spring-slider" systems. These systems involve three microscopic parameters that must be determined by calibration: the microscopic densification pressure σ_{μ}^{d}, the microscopic tangential modulus E_{μ}^{t} and the maximum permanent deformation of the beam bonds $\varepsilon_{\mu}^{p\,max}$. To simplify the calibration of these parameters, a parametric study was performed to investigate the influence of each parameter on the resulting macroscopic mechanical behavior. The validation of the proposed model was first preformed in quasi-statics to better analyze the potential problems in the absence of dynamic effects. A quasi-static compression of a sphere was simulated to validate the model at the macroscopic scale. Then, to investigate the effectiveness of this model at the microscopic scale, it was applied to simulate quasi-static Vickers microindentation tests. The obtained results are in good agreement with experimental results taken from the literature [JI 07, YOS 05]. This proves the ability of the proposed model to simulate the silica glass densification under a quasi-static loading. Therefore, it can be concluded that the modeling assumption (shear effects are neglected) used to develop this model is valid. To validate the model in fast dynamics (scope of the book), it was applied to simulate high-velocity impact tests of silica glass plates. As shown, the different features of the material response under highly dynamic loadings are correctly reproduced [SUG 81]. Furthermore, quantitatively, the obtained results would be acceptable, compared to other numerical results from the literature [JI 07, MIC 07].

To model the silica glass cracking behavior, the virial-stress-based fracture model proposed in the first book of this set [JEB 15] was retained. This model allows for better simulation of the brittle cracking mechanisms at both the macroscopic and microscopic scales. To verify the ability of this model to correctly reproduce the silica glass fracture behavior, it was applied to simulate one of the most challenging cracking problems of this material: fracture of silica glass under the Vickers indenter. The obtained results are very promising, compared to other experimental results [ARO 79].

The proposed model of the silica glass mechanical behavior as well as the retained fracture model will be used in Chapter 6 to simulate the LSP of this material.

6

Simulation of Laser Shock Processing of Silica Glass

6.1. Introduction

Nowadays, laser shock processing (LSP) of materials has become widely used in various engineering areas to enhance near-surface mechanical properties of metals, to remove matter by cutting or drilling, to weld metals up to several centimeters in thickness, to manufacture complex shapes by powder melting, to harden or texture surfaces at the microscopic or mesoscopic scale and still others remaining to be discovered. The ever-growing use of the LSP process has spawned a need for more in-depth studies to deal with emerging issues. To meet such needs, several experimental studies have been performed over the last decades [LES 06, BOU 98, GRA 78, COT 89, FAB 98, KAN 98, BER 11]. Nevertheless, laser experiments are generally very costly and ineffective in terms of providing insightful and complete information of what happens inside the studied specimen. Much of the LSP information in which the designer is interested is very difficult to obtain experimentally. Figure 6.1 shows the typical experimental outcomes that can be obtained, in a relatively straightforward manner, from an LSP test. These results were obtained at LULI[1] using thin silica glass plates by a research team, with whom we collaborate, composed of J. C. Sangleboeuf, J. P. Guin, M. Nivard, D. Loison, L. Berthe, M. Boustie, E. Lescoute, A. Sollier and G. Sesson. Despite the great experimental advances, high-performance experimental means would

1 *Laboratoire d'Utilisation des Lasers Intenses* (École polytechnique, Paris, France).

not be sufficient to capture some relevant small-scale effects that can greatly influence the LSP results. This makes some emerging LSP challenges beyond the reach of the current experimental advances. Therefore, it is necessary to imply the numerical simulation to better analyze these challenges and to overcome the experimental difficulties.

As seen in Chapter 4, LSP of materials mainly involves two steps. The first step consists of converting the deposited laser energy into laser pressure. As a second step, the laser-induced pressure starts to propagate within the irradiated material. To model this last step, which is of interest to the mechanical engineering researchers, it is necessary to assess the temporal and spatial pressure profiles resulting from the first step. Several techniques have been proposed in the literature to approximate the temporal pressure profile. However, these techniques are generally based on simplifying assumptions which can result in poor or even unrealistic temporal pressure approximation. In addition, the spatial distribution of the laser pressure is generally very irregular and unpredictable. Therefore, further research effort is required to get over these difficulties, before being able to study numerically the complex LSP phenomena induced by the propagation of the laser pressure within the studied specimen. In this context, an interesting path to be explored consists of applying the developed discrete element method-constrained natural element method (DEM-CNEM) coupling approach as an inverse technique to obtain information on the laser pressure from the final experimental results (Figure 6.1). This information would then be used to enrich the pressure profile computed by a specialized software. After obtaining accurate temporal and spatial profiles of the laser pressure, it would be possible to investigate numerically some complex LSP phenomena that cannot be easily examined with experimental techniques. Implementation of this idea requires a preliminary study to test its reliability and to ensure that the developed coupling approach can correctly predict important mechanisms usually encountered in LSP applications. This is the objective of the current chapter which aims to study qualitatively the mechanical part of LSP of silica glass. Using pressure profiles inspired by those obtained using specialized software packages, the following sections try to analyze qualitatively the important LSP mechanisms caused by pressure propagation within the considered silica glass specimen. A brief description of the LSP test considered in this study is given in the next section.

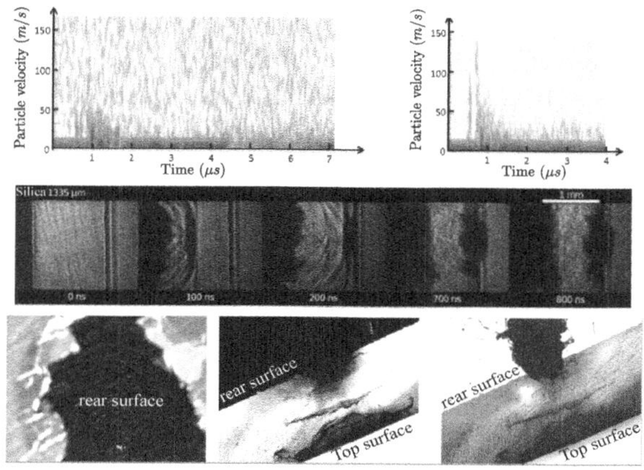

a) Laser intensity $I_L = 1.640\,PW/cm^2$; top: particle velocity at the rear surface; middle: shock wave propagation; bottom: silica glass fracture

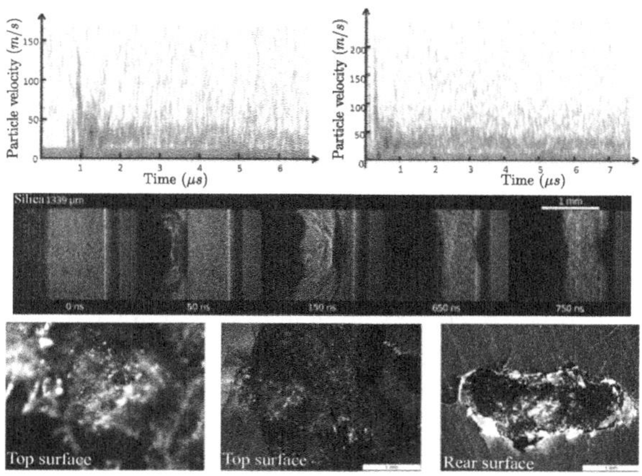

b) Laser intensity $I_L = 1.546\,PW/cm^2$; top: particle velocity at the rear surface; middle: shock wave propagation; bottom: silica glass fracture

Figure 6.1. *Laser shock processing of thin silica glass plates using different laser intensities. For a color version of the figure, see www.iste.co.uk/jebahi/discrete2.zip*

6.2. LSP test

A silica glass plate having a square base of $20 \times 20\,mm^2$ and thickness of $5\,mm$ is considered in this chapter. To convert the laser energy into pressure, two distinct ablation modes exist: direct ablation and confined ablation [DAH 08]. In direct ablation, the irradiated surface is directly exposed to the laser beam, which leads to free expansion of the generated plasma. This ablation mode generally results in low laser-induced pressures. In order to better investigate the ability of the DEM-CNEM approach to reproduce the different silica glass specificities under laser shock, high laser pressure should be used. Therefore, confined ablation is considered in this chapter. In this ablation mode, the generated plasma is confined during its expansion by a transparent material (e.g. water) placed in close contact with the irradiated surface. Since silica glass is transparent to the laser beam, an aluminum coating layer is deposited between the irradiated surface and the confining transparent material (Figure 6.2). In summary, the top surface of the silica glass plate is assumed to be covered with two overlays: an opaque overlay (opaque to the laser beam) directly on the plate top surface, and over this a transparent overlay (transparent to the laser beam). When the laser beam is directed onto this surface, it passes through the transparent overlay and strikes the opaque one. This causes the opaque overlay to immediately vaporize. Then, the generated vapor absorbs the incoming laser energy, rapidly heats and expands against the transparent overlay and the plate top surface. The transparent overlay is only used to confine the thermally expanding vapor against the top surface. Consequently, the pressure can rise much higher and longer (in duration) than it would if the transparent overlay was absent (i.e. in direct ablation). The sudden rise of the pressure against the top surface leads to a shock wave which will propagate in the plate thickness. The laser beam is assumed to interact only with the aluminum (opaque) overlay. Furthermore, the important thermal effects appearing in this layer are assumed to propagate slowly, compared to the shock pressure. Therefore, the thermal behavior of silica glass is neglected in the present work.

Concerning the laser loading properties, the silica glass plate is assumed to be subjected to a high-power laser beam of $3\,mm$ diameter and several Joules energy-per-pulse, so that the aluminum overlay is hit by a powerful laser density. As mentioned in Chapter 4, several software packages specialized in laser–matter interaction can be found in the literature [MAI 08, COL 06]. These are mainly used to evaluate the laser-induced pressure. This part is beyond the scope of this chapter which studies the LSP processing from a mechanical point of view. A temporal pressure profile close to those resulting from these software packages is used to model the laser-induced pressure.

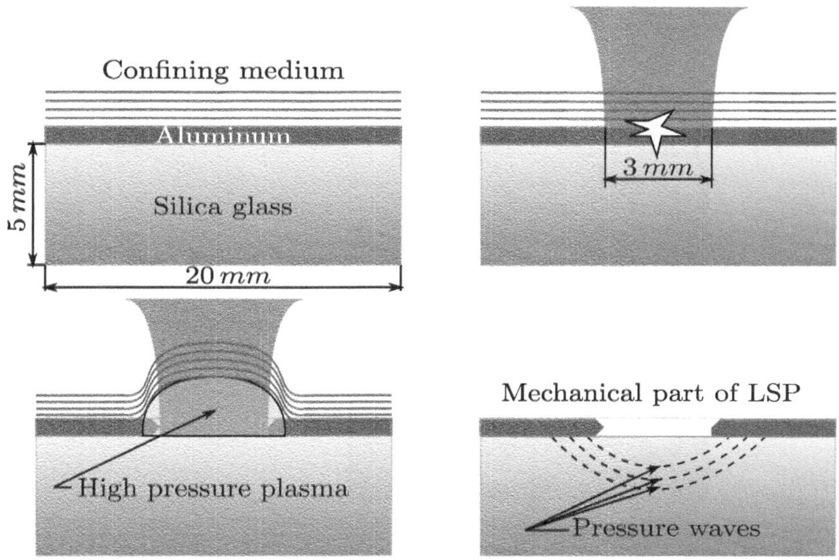

Figure 6.2. *Description of the LSP test*

6.3. LSP model

The LSP test presented in the previous section involves several complex multi-scale phenomena. The region closely surrounding the laser beam is characterized by very complex effects that require a very fine-scale analysis. This region must be modeled by DEM using a large number of discrete elements. The rest of the silica glass plate undergoes only structural effects, due to the transversal propagation of the laser-induced waves. Therefore, to avoid crippling time computation, the regions beyond the laser-affected zone must be modeled by CNEM. Figures 6.3 and 6.4 show the geometric model used to simulate the LSP test. The discrete element method (DEM) is only applied in a region of about $6\,mm$ diameter, located in the middle of the considered silica glass plate. The constrained natural element method (CNEM) is applied in the remaining estate. The DEM and CNEM models overlap in a small region having hollow-cylindrical shape of $4\,mm$ internal diameter and $6\,mm$ external diameter (Figure 6.4). Although the fine-scale effects to be studied are only localized in the DEM subdomain, it is necessary to consider the regions (modeled by CNEM) beyond this subdomain to move away the boundary condition effects. This prevents spurious wave reflection at the lateral border of the DEM subdomain, and then allows the shock front to correctly propagate in the loading direction.

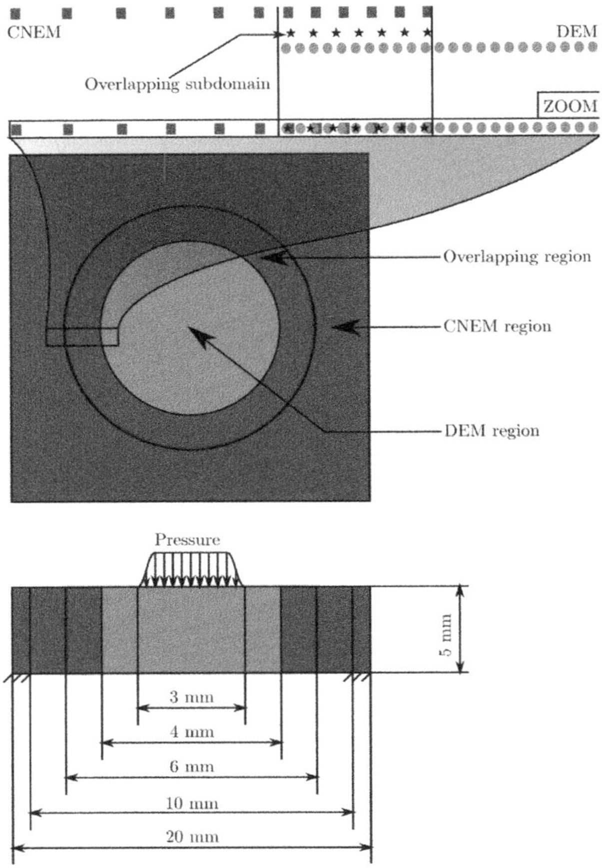

Figure 6.3. *Geometric coupling model*

To model the laser-induced pressure, the shock surface is subjected to a pressure loading, of which the temporal profile is close to those obtained in the literature using a laser–matter interaction software. As shown in Chapter 4, when the amplitude and duration of the laser-induced pressure is sufficient, the multiple spalling phenomenon can take place (Figure 4.12). This phenomenon is very difficult to reproduce with traditional numerical methods. In an attempt to reproduce this phenomenon with the developed DEM-CNEM approach, relatively high pressure with long duration is applied on the shock surface (Figure 6.5(a)). The diameter of this surface is assumed to be the same as that of the laser beam ($3\,mm$). The spatial distribution of the laser-induced pressure is generally very irregular and unpredictable. In this qualitative study, a simple pressure distribution is applied on the shock surface (Figure 6.5(b)), a plateau with gradual decrease at the border to reduce the shear stresses in this zone.

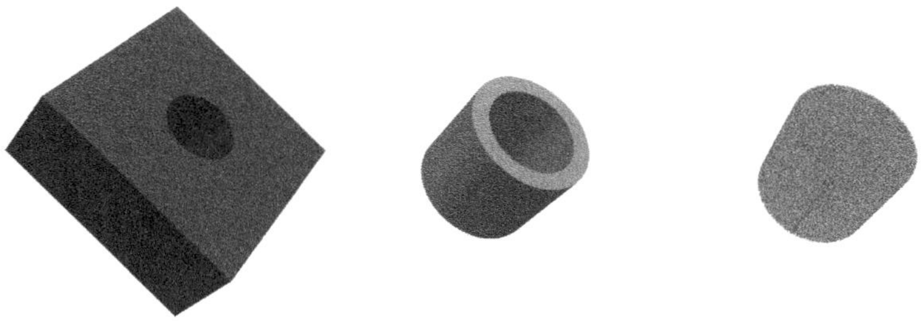

a) CNEM subdomain b) Overlapping subdomain c) DEM subdomain

Figure 6.4. *3D visualization of the different coupling subdomains*

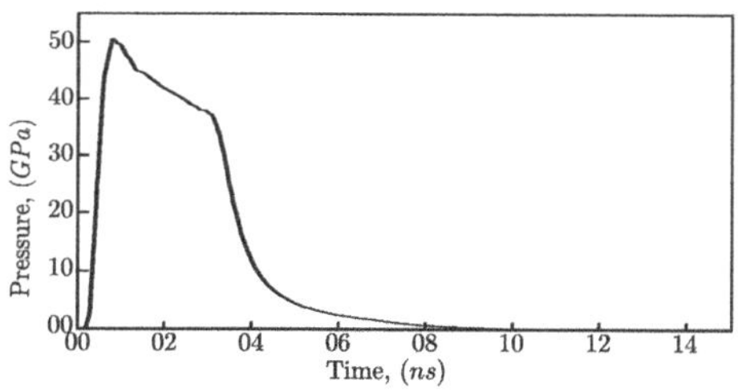

a) Temporal pressure profile

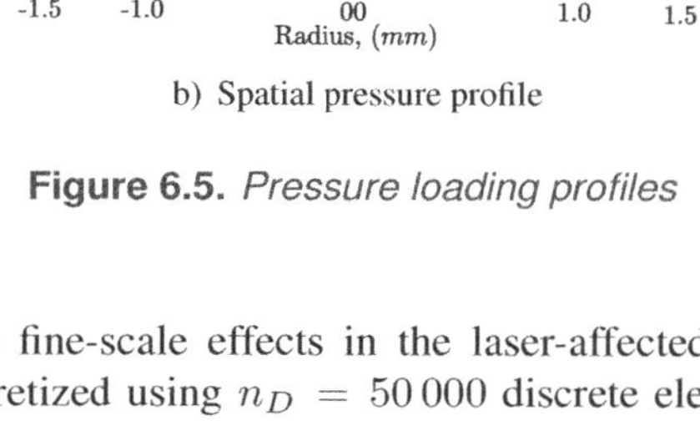

b) Spatial pressure profile

Figure 6.5. *Pressure loading profiles*

To capture the fine-scale effects in the laser-affected region, the DEM subdomain is discretized using $n_D = 50\,000$ discrete elements (the average radius of the discrete elements is about $r_{particle}^{mean} = 75\,\mu m \approx h_D$). If the entire studied domain was discretized using the same $r_{particle}^{mean}$, the number of discrete elements would exceed $n_D = 500\,000$. This would require several weeks of DEM computation just to prepare the DEM subdomain, hence the importance of using the coupling approach. The CNEM subdomain is coarsely discretized. Far from the bridging region, the characteristic length of the CNEM discretization is chosen to be $h_C = 10\,h_D$. This parameter is gradually decreased when getting close to the bridging region Ω_B. This is required to ensure a sufficient number of CNEM nodes in this region. The part of the CNEM subdomain belonging to Ω_B is discretized such that h_C is approximately five times larger than h_D ($h_C = 5\,h_D$). Based on the results of the parametric study given in section 3.4, the discretization of the third subdomain specific to the bridging region is performed such that the associated characteristic length is $h_B = 2\,h_D$. Concerning the junction parameter l, it is shown that its optimal value l_{opt} is roughly of the same order of magnitude as h_B. Therefore, this parameter is fixed at $l = h_B$. The discrete and continuum energies in the overlapping region are weighted using continuous differential weight functions (Figure 1.5(c)). To validate this choice of coupling parameters, a preliminary dynamic simulation was performed, using only elastic behavior for both DEM and CNEM models. In this simulation, a pressure loading with very steep slope and moderate magnitude of $1GPa$ (to remain within the elastic domain) is applied on the shock surface. Figure 6.6 compares the free surface (rear surface) displacements with respect to time obtained from DEM-CNEM computation and fully CNEM computation. The coupling result is fairly close to the CNEM result. This guarantees the validity of the different coupling parameters that are chosen in this section.

To model the complex mechanical behavior of silica glass in the DEM region, the beam-based model developed in Chapter 5 is used in this region. However, only elastic behavior is considered in the CNEM subdomain, because CNEM is applied far from the laser-affected zone. Since the magnitude of the pressure loading (Figure 6.5) is much higher than $P_{nle} = 3\,GPa$ (the pressure beyond which the elastic behavior becomes linear), the silica glass nonlinear elastic behavior is not considered in the present qualitative study. This makes the application of the DEM-CNEM coupling approach easier and considerably reduces the computation time. Concerning the silica glass fracture, the virial-stress-based cracking model developed in the first book of this set of books [JEB 15] is used to model the cracking behavior of this material in the DEM region. The present qualitative study mainly focuses on the spalling phenomenon which is very difficult to reproduce numerically. Therefore, calibration of this model is done using the spalling fracture strength of the material. As seen in Chapter 4, this silica glass property is difficult to measure experimentally and can exceed $2.5\,GPa$. In this qualitative study, it is fixed at $2.5\,GPa$.

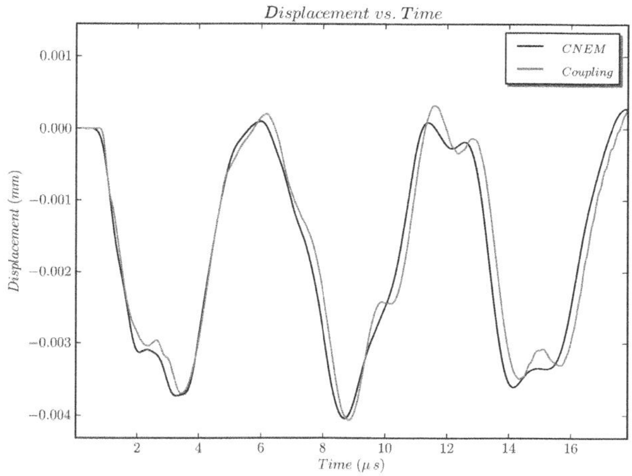

Figure 6.6. *Time evolution of the free surface (rear surface) displacements obtained using DEM-CNEM coupling approach and CNEM approach*

6.4. Results

This section presents the simulation results associated with the numerical model presented in the previous section. Since the phenomena of interest are localized in the DEM subdomain, only DEM results are presented in hereafter. Figure 6.7 illustrates the propagation of the pressure waves in the DEM subdomain. A compression pressure wave propagates in the plate thickness, and is immediately followed by a release wave due to the rapid decrease in the pressure loading (Figure 6.5). When reaching the rear (free) surface, the compression peak pressure measures only $P = 8\,GPa$. Two physically based damping mechanisms can explain this pressure decrease, as discussed in section 4.2.2.4. The first mechanism is known as elastoplastic damping. Beyond the $HEL = 10\,GPa$, the shock front is split into two parts: elastic shock front and densification (plastic) shock front traveling at, respectively, the longitudinal sound velocity C_l and the shock velocity V_s. Based on Table 5.4, the longitudinal sound velocity C_l is approximately $4950\,ms^{-1}$. Furthermore, an approximate value of V_s can be obtained using the silica glass shock polar presented in Figure 4.19: $V_s \approx 3300\, ms^{-1}$. During rarefaction (unloading), the elastic part of the release wave follows the shock front at C_l. Since in the present case $C_l > V_s$, the elastic release part can overtake the densification shock front. This results in reducing the peak pressure. The second damping mechanism is the hydrodynamic damping. Since the silica glass Hugoniot curve is concave (Figure 5.17), the higher amplitude regions of the pressure waves travel faster than the lower amplitude regions. Therefore, the head of the release wave (corresponding to the peak pressure) travels faster than the shock front. This reduces progressively the pulse duration to zero. After it is zero, the peak pressure begins to decline, which can explain the pressure drop to $P = 8\, GPa$ at the rear surface. In addition to these two physically based mechanisms, another numerical mechanism due to the discretization of the problem domain can also explain the pressure decrease. As explained in section 1.2.2, there is a cutoff frequency, depending on the DEM discretization, beyond which the frequency waves are filtered. This can cause a reduction in the pressure that propagates within the DEM subdomain. To alleviate the numerical damping with respect to the physically based ones, the discretization of the DEM subdomain was preformed so as to support the major part of the pressure loading.

Figure 6.8 presents the time evolution of the particle velocity at the rear surface. No dual wave is observed on the profile of particle velocity. This means that only an elastic wave reached the rear surface. At this surface, the maximum particle velocity is around $V_p = 800\, ms^{-1}$. Based on the same Figure 6.8, the velocity of the shock front (elastic wave) can also be deduced. This velocity is about $V_s = 4961\, ms^{-1}$, which is in good agreement with the experimental and numerical results presented in Table 5.4. The particle velocity profile can also be used to get information on the fracture of the rear surface. As seen in Figure 6.8, the particle velocity has stabilized at around $117\, ms^{-1}$ without reversing direction. This means that there is fracture of the material and formation of spalls. The great difference between the maximum and stabilized particle velocities indicates that a thick spall is thought to be formed. Figure 6.9 presents the damage state of the DEM subdomain due to the shock wave propagation. Two consecutive spalls were formed at the rear surface (free surface). It should be noted that the multiple spalling phenomenon is very difficult to capture numerically. The spalling result obtained in this work is then very important and proves the effectiveness of the DEM model. Indeed, such a result indicates that this model can correctly reproduce the reflection mechanisms of the incoming compression pressure at the rear surface. In addition to the rear surface spalling, there is also damage of the front surface (shock surface), due to the fast pressure loading decrease (release wave). This is in agreement with the experimental observations.

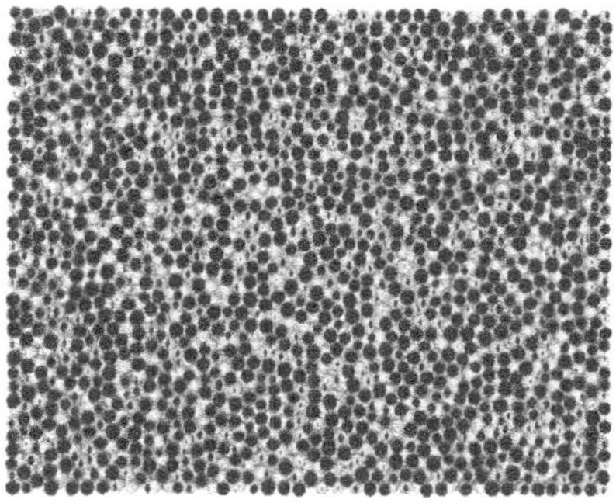

a) $t < 0\,\mu s$, $P^{comp} = 50\,GPa$

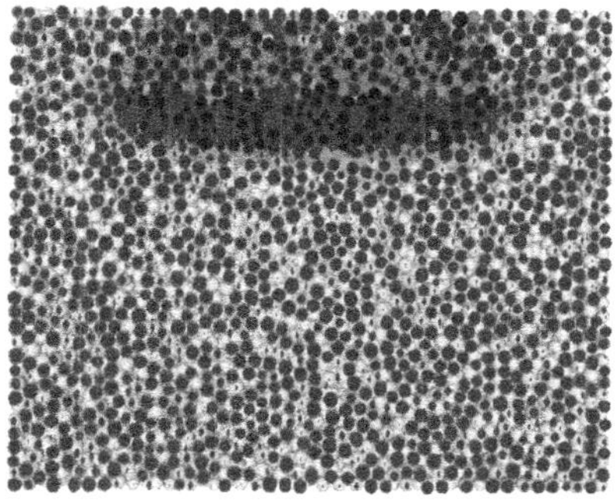

b) $t = 0.15\,\mu s$, $P^{comp} = 35\,GPa$

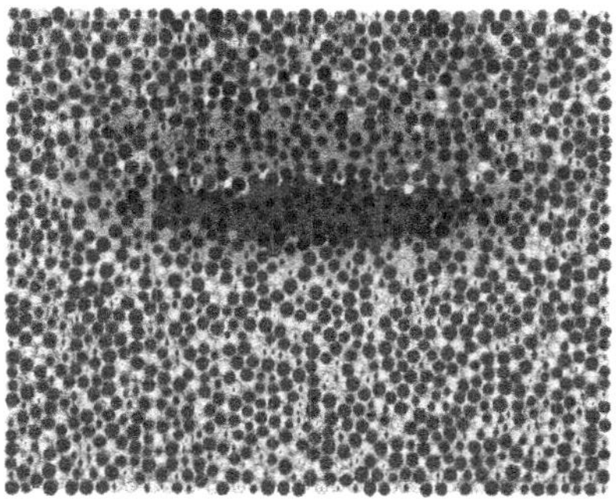

c) $t = 0.31\,\mu s$, $P^{comp} = 20\,GPa$

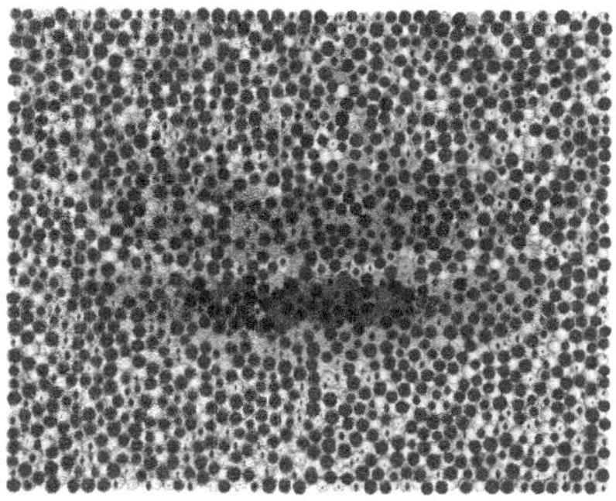

d) $t = 0.47\,\mu s$, $P^{comp} = 15\,GPa$

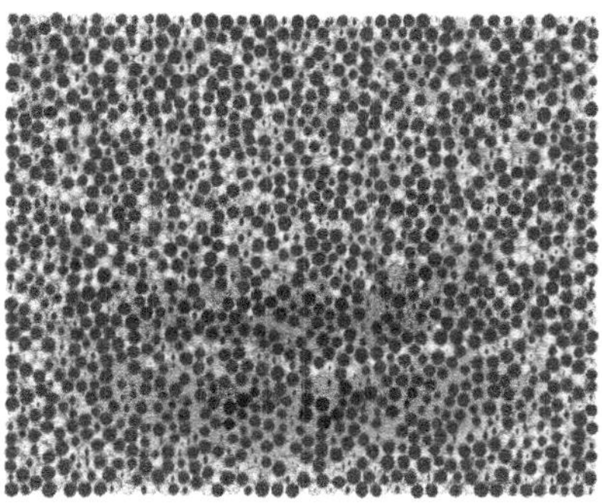

e) $t = 0.63\,\mu s$, $P^{comp} = 10\,GPa$

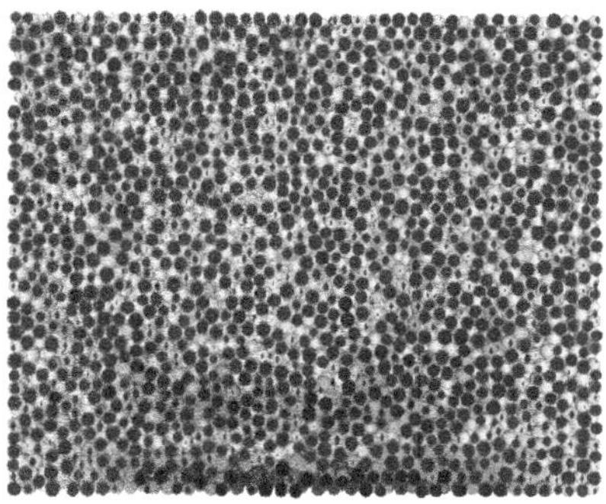

f) $t = 0.79\,\mu s$, $P^{comp} = 8\,GPa$

Figure 6.7. *Pressure wave propagation in the DEM region; a compression wave immediately followed by a release wave. For a color version of the figure, see www.iste.co.uk/jebahi/discrete2.zip*

The different numerical results obtained in this section are qualitatively in agreement with the theory of shock wave propagation in materials (presented in Chapter 4) and with the experimental observations on the LSP of materials [LES 06, PHI 02].

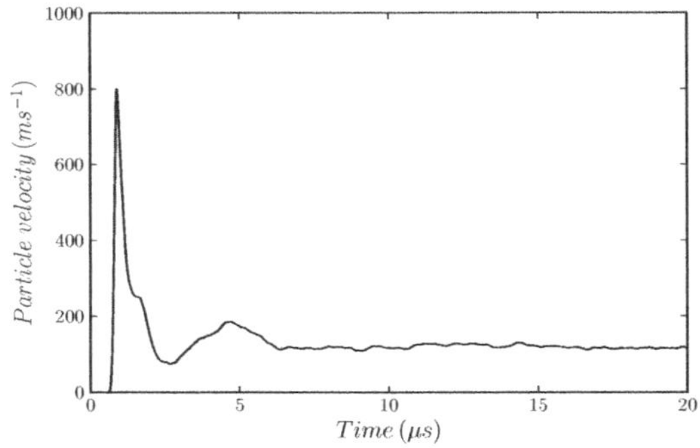

Figure 6.8. *Particle velocity measured at the free surface (rear surface)*

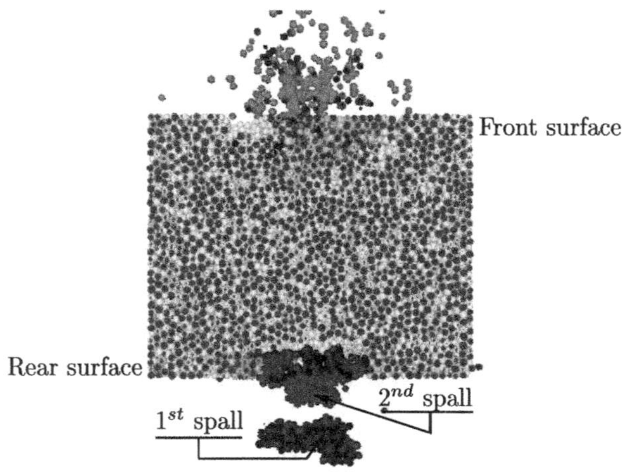

Figure 6.9. *Damage state of the DEM subdomain due to the shock wave propagation. For a color version of the figure, see www.iste.co.uk/jebahi/discrete2.zip*

6.5. Conclusion

In this chapter, a qualitative study of the LSP of silica glass was performed, using the DEM-CNEM coupling approach. To reduce the computation time, only a small region surrounding the laser beam is modeled by DEM, whereas CNEM is applied in the rest of the studied domain to move away the boundary conditions effects. To correctly model the silica glass mechanical behavior in the DEM subdomain where complex phenomena can take place, the beam-based model developed in Chapter 5 was applied in this subdomain. In addition, the virial-stress-based cracking model developed in the first book of this set [JEB 15] was also used to model the silica glass cracking behavior due to the propagation of the pressure shock wave in the DEM subdomain. Since CNEM is applied far from the high-loaded regions, only elastic behavior was applied in the associated subdomain. Concerning the laser-induced pressure, the shock surface was subjected to a temporal pressure profile close to those obtained using a specialized laser–matter interaction software. To capture the multiple spalling phenomenon which is very challenging for traditional numerical methods, relatively high peak pressure with long duration was assumed in this qualitative study. Furthermore, a simple spatial distribution was used to apply the pressure loading on the shock surface: a plateau with gradual decrease at the border to reduce the shearing effects in this zone. Since the peak pressure is much higher than P_{nle}, the nonlinear elastic behavior was also neglected in this study.

The obtained qualitative results are in agreement with the experimental observations on the LSP of materials. This is very promising to move on to the quantitative study of this process. The proposed coupling approach will then be used as an inverse technique to obtain accurate temporal and spatial pressure profiles, from measurable experimental results. By doing so, it would be possible to simulate quantitatively such process, and to assist the interpretation or even the discovery of new LSP phenomena.

Conclusion

The first book of this set [JEB 15] proposed a discrete element method (DEM) variation allowing for correct discrete element modeling of continuous materials. This approach was successfully applied to investigate several challenging problems that cannot be easily treated by continuum methods. Nowadays, it presents an alternative way to study complex phenomena requiring very small-scale analysis. The major drawback of this method is that it is very costly compared to continuum methods and the computation time can quickly become overwhelming, especially in the case of large-studied domains. However, in modern science and engineering, real materials usually exhibit multi-scale phenomena that require on one scale a very accurate description to capture the complex effects at this scale, and on another scale a coarse description to avoid prohibitively large computation. The ever-accelerating scientific advances have given rise to numerous interesting problems that need multi-scale modeling. Of particular interest are the fast dynamic problems which generally involve strongly dependent multi-scale effects. Due to their complexity, these problems are very challenging for traditional numerical methods. The proposed DEM can, however, give solutions for several outstanding difficulties related to these problems. A common feature of such problems is that the regions requiring a fine-scale analysis by DEM are generally small with respect to the entire problem domain. Therefore, with a view of expanding the scope of DEM while alleviating its drawbacks, this book developed a concurrent discrete-continuum coupling approach adapted for highly dynamic problems between DEM and a continuum method. This is the first objective of the this book. As an application of this coupling approach, it was applied to study qualitatively the laser shock processing (LSP) of silica glass. This is the second goal of the work. To achieve these two requests, this book is divided, after a general introduction, into two parts. Part 1 dealt with the development of the discrete-continuum coupling approach. This part is composed of three chapters:

– Chapter 1 brought light to some important aspects related to the concurrent coupling approach used to couple discrete and continuum methods for strongly dependent multi-scale phenomena. First, the major difficulties related to the application of this approach to couple dissimilar methods for highly dynamic studies were briefly outlined. These difficulties are mainly due to the different mechanics and the different analysis scales between the coupled methods. Then, the techniques most commonly used to overcome these difficulties were reviewed. Among these techniques, the bridging domain technique was retained to couple DEM with a continuum method. Using this technique, spurious wave reflections can be avoided without any additional filtering or damping.

– Chapter 2 focused on the choice of the appropriate continuum method to be coupled with DEM for highly dynamic problems. To simplify this task, the continuum methods were classified into two classes: grid-based methods and meshless methods. Comparison of these classes with respect to the expectations of this work allowed for retaining the meshless methods. These methods, in turn, can be divided into approximation methods and interpolation methods. From these categories, the interpolation methods offer better solutions for difficulties related to the imposition of the boundary conditions. Therefore, they are best adapted for coupling with DEM. Among the interpolation meshless methods, the constrained natural element method (CNEM) is the most recent and includes practically all the advantages of this category. Therefore, it was retained to be part of the discrete-continuum coupling. The main features of this method were then recalled to better understand the development of the coupling formulation.

– Chapter 3 detailed how DEM and CNEM were coupled. Based on the bridging domain technique, a three-dimensional (3D) DEM-CNEM coupling approach was developed. Several coupling parameters are involved in this approach. Therefore, to simplify their setting in practice, a parametric study of these parameters was performed. This study allowed for retaining several recommendations simplifying the setting of these parameters for complex applications. Finally, this new coupling method was validated using several dynamic reference tests. From an information technology (IT) point of view, since the GranOO libraries are written in C++ language, the DEM side coupling algorithm was implemented in the same language. The NESSY libraries are written in C++ language and interfaced with Python moduli. Therefore, the CNEM side coupling algorithm has been implemented in Python language. The communication between the two codes (GranOO and

NESSY) is ensured by a DC-Interface developed in this work in Python language. DC-Interface communicates directly with NESSY (same language) and via an Inter Process Communication (IPC) with GranOO.

The developed DEM-CNEM coupling approach was applied to study qualitatively one of the most challenging problems in fast dynamics: LSP of silica glass. This is the subject of Part 2 which is also divided into three chapters:

– Chapter 4 introduced the main scientific advances in the different disciplines interacting to accomplish the second goal of this work. First, some basic concepts related to the theory of laser–matter interaction were given. The physics behind this theory can be divided into two steps: generation of shock wave by laser ablation and shock wave propagation within the irradiated material. The first step is not within the scope of this work which only focuses on the mechanical part of the laser shock processing of silica glass. In this qualitative study, the temporal pressure profile generated by the laser–matter interaction (in the first step) was assumed to be given and was directly applied on the shock surface. Second, to better characterize the silica glass response under high pressures generated by laser shock, the main experimental works on the complex mechanical behavior of silica glass were reviewed. In light of this review, a numerical model of this material was developed in Chapter 5.

– Chapter 5 was devoted to the modeling of the silica glass mechanical behavior by the DEM-CNEM coupling approach. In this approach, CNEM would be applied in the regions far from the complex LSP effects. Therefore, only elastic behavior is considered in these regions. On the contrary, DEM would be applied in the laser-affected zones. To correctly model the silica glass response in these zones, a new silica glass model adapted for DEM was proposed. This model is based on an assumption that the shearing effects (matter of controversy in the literature) are negligible, compared to pressure effects. To model the nonlinear elastic behavior in the region of low pressures, empirical relationships were proposed to translate the macroscopic variations of the silica glass mechanical properties into microscopic variations of the cohesive beam parameters. To model the silica glass densification, the original elastic cohesive beams between discrete elements are connected in series with "spring-slider" systems. These systems involve three microscopic parameters that must be determined by calibration. To simplify this task, the influence of these parameters on the macroscopic mechanical behavior was studied. The validation of the proposed model was first validated in quasi-statics to better analyze the potential problems in the absence of the dynamic effects. A quasi-static compression of a sphere was simulated to validate the model at the macroscopic scale. Then, a Vickers microindentation was simulated to investigate the effectiveness of the model at the microscopic scale. The obtained results are in good agreement with experimental results taken from the literature [JI 07, YOS 05]. This proves the validity of the modeling assumption that the shearing effects are negligible, compared to the hydrostatic pressure effects. To validate the proposed model in fast dynamics (scope of this work), it was applied to simulate high-velocity impact tests of silica glass plates. As shown, the different features characterizing the silica glass response under highly dynamic loading were correctly reproduced. Furthermore, quantitatively, the obtained results would be acceptable compared to other numerical results [JI 07, MIC 07]. To model the silica glass cracking behavior, the virial-stress-based fracture model proposed in the first book of this set of books [JEB 15] was used. To verify the ability of this model to correctly reproduce the silica glass fracture behavior, it was applied to simulate one of the most challenging cracking problems of this material: fracture of silica glass under Vickers indenter. The obtained results are in good agreement with other experimental results [ARO 79].

– Chapter 6 focused on the application of the DEM-CNEM coupling approach to study qualitatively the LSP of silica glass. To reduce the computation time, only a small region surrounding the laser beam was modeled by DEM, whereas, CNEM was applied in the rest of the studied domain to move away the boundary condition effects. To correctly model the silica glass mechanical behavior in the DEM subdomain, the beam-based model developed in this work was applied. In addition, the virial-stress-based cracking model developed in the first book of this set of books [JEB 15] was also used to model the silica glass cracking behavior in the DEM subdomain. On the contrary, only elastic behavior was applied in the CNEM subdomain. Concerning the laser-induced pressure, the shock surface was subjected to a temporal pressure profile close to those obtained using a specialized laser–matter interaction software. To capture the multiple spalling phenomenon which is very challenging for traditional numerical methods, relatively high-peak pressure with long duration was assumed in this qualitative study. Furthermore, a simple spatial distribution was used to apply the pressure loading on the shock surface. Since the peak pressure is much higher than P_{nle}, the nonlinear elastic behavior was also neglected in this study. The obtained qualitative results are in agreement with the experimental observations on the LSP of materials. This is very promising to move on to the quantitative study of this process.

With the development of the DEM-CNEM coupling approach, several complex highly dynamic problems have become affordable. As a future work, this approach will be used as an inverse technique to obtain accurate temporal and spatial profiles of the laser-induced pressure, from measurable experimental results. By doing so, it would be possible to simulate quantitatively the LSP of materials, and to assist the interpretation or even the discovery of new LSP phenomena. All the developments performed in this work were implemented in GranOO workbench by adding plug-ins written in C++ language using the advanced Oject Oriented (OO) architecture. The third book of this set aims to present this workbench from an IT point of view.

Bibliography

[ABR 98] ABRAHAM F., BROUGHTON J., BERNSTEIN N. *et al.*, "Spanning the continuum to quantum length scales in a dynamic simulation of brittle fracture", *Europhysics Letters*, vol. 44, pp. 783–787, 1998.

[ADE 76] ADELMAN S., DOLL J., "Generalized Langevin equation approach for atom/solid-surface scattering: general formulation for classical scattering off harmonic solids", *Journal of Chemical Physics*, vol. 64, pp. 2375–2388, 1976.

[AKB 12] AKBARI R.A., KERFRIDEN P., RABCZUK T. *et al.*, "An adaptive multiscale method for fracture based on concurrent – hierarchical hybrid modelling", *Proceedings of the 20th UK Conference of the Association for Computational Mechanics in Engineering*, 2012.

[AND 68] ANDERHOLM N.C., "Laser generated pressure waves", *Bulletin of the American Physical Society*, vol. 13, p. 388, 1968.

[AND 12a] ANDRÉ D., Modélisation par éléments discrets des phases d'ébauchage et de doucissage de la silice, PhD Thesis, University of Bordeaux 1, 2012.

[AND 12b] ANDRÉ D., IORDANOFF I., CHARLES J.L. *et al.*, "Discrete element method to simulate continuous material by using the cohesive beam model", *Computer Methods in Applied Mechanics and Engineering*, vol. 213–216, pp. 113–125, 2012.

[AND 13] ANDRÉ D., JEBAHI M., IORDANOFF I. *et al.*, "Using the discrete element method to simulate brittle fracture in the indentation of a silica glass with a blunt indenter", *Computer Methods in Applied Mechanics and Engineering*, vol. 265, pp. 136–147, 2013.

[AND 14] ANDRÉ D., CHARLES J.L., IORDANOFF I. *et al.*, "The GranOO workbench, a new tool for developing discrete element simulations, and its application to tribological problems", *Advances in Engineering Software*, vol. 74, no. 0, pp. 40–48, 2014.

[ANT 02] ANTOUN T., SEAMAN L., CURRAN D.R. *et al.*, *Spall Fracture*, Springer, 2002.

[ARO 79] ARORA A., MARSHALL D., LAWN B. *et al.*, "Indentation deformation/fracture of normal and anomalous glasses", *Journal of Non-Crystalline Solids*, vol. 31, pp. 415–428, 1979.

[ATL 98] ATLURI S.N., ZHU T., "A new meshless local Petrov-Galerkin (MPLG) approach in computational mechanics", *Computational Mechanics*, vol. 22, pp. 117–127, 1998.

[AUB 10] AUBERTIN P., RÉTHORÉ J., DE BORST R., "A coupled molecular dynamics and extended finite element method for dynamic crack propagation", *International Journal for Numerical Methods in Engineering*, vol. 81, pp. 72–88, 2010.

[BAU 08] BAUMAN P.L., BEN DHIA H., ELKHODJA N. *et al.*, "On the application of the Arlequin method to the coupling of article and continuum models", *Computational Mechanics*, vol. 42, pp. 511–530, 2008.

[BAU 09] BAUMAN P.T., ODEN J.T., PRUDHOMME S., "Adaptive multiscale modeling of polymeric materials with Arlequin coupling and goals algorithms", *Computer Methods in Applied Mechanics and Engineering*, vol. 198, nos. 5–8, pp. 799–818, 2009.

[BEI 96] BEISSEL S., BELYTSCHKO T., "Nodal integration of the element-free Galerkin method", *Computer Methods in Applied Mechanics and Engineering*, vol. 139, pp. 49–74, 1996.

[BEL 94] BELYTSCHKO T., LU Y.Y., GU L., "Element-free Galerkin methods", *International Journal for Numerical Methods in Engineering*, vol. 37, pp. 229–256, 1994.

[BEL 96] BELYTSCHKO T., KRONGAUZ Y., ORGAN D. *et al.*, "Meshless methods: an overview and recent developments", *Computer Methods in Applied Mechanics and Engineering*, vol. 139, pp. 3–47, 1996.

[BEL 97] BELIKOV V.V., IVANOV V.D., KONTOROVICH V.K. *et al.*, "The non-Sibsonian interpolation: a new method of interpolation of the values of a function on an arbitrary set of points", *Computational Mathematics and Mathematical Physics*, vol. 37, no. 1, pp. 9–15, 1997.

[BEL 03] BELYTSCHKO T., XIAO S.P., "Coupling methods for continuum model with molecular model", *International Journal for Multiscale Computational Engineering*, vol. 1, no. 1, pp. 115–126, 2003.

[BEN 92] BENSON D.J., "Computational methods in Lagrangian and Eulerian hydrocodes", *Computer Methods in Applied Mechanics and Engineering*, vol. 99, nos. 2–3, pp. 235–394, 1992.

[BEN 95] BENZ W., ASPHAUG E., "Simulation of brittle solids using smooth particle hydrodynamics", *Computer Physics Communications*, vol. 87, nos. 1–2, pp. 253–265, 1995.

[BEN 98] BEN DHIA H., "Problèmes mécanique multi-échelles: la méthode Arlequin", *Comptes rendus de l 'Académie des sciences – Analyse numérique*, vol. 326, no. 12, pp. 899–904, 1998.

[BEN 01] BEN DHIA H., RATEAU G., "Analyse mathématique de la méthode Arlequin mixte", *Comptes rendus de l 'Académie des sciences – Mécanique des solides et des stuctures*, vol. 332, no. 7, pp. 649–654, 2001.

[BEN 05] BEN DHIA H., RATEAU G., "The Arlequin method as a flexible engineering design tool", *International Journal for Numerical Methods in Engineering*, vol. 62, pp. 1442–1462, 2005.

[BEN 08] BEN DHIA H., "Further insights by theoretical investigations of the multiscale Arlequin method", *International Journal for Multiscale Computational Engineering*, vol. 60, no. 3, pp. 215–232, 2008.

[BER 11] BERTHE L., ARRIGONI M., BOUSTIE M. *et al.*, "State-of-the-art laser adhesion test (LASAT)", *Nondestructive Testing and Evaluation*, vol. 26, no. 3, pp. 303–317, 2011.

[BOU 98] BOURNE N., MILLETT J., ROSENBERG Z. *et al.*, "On the shock induced failure of brittle solids", *Journal of the Mechanics and Physics of Solids*, vol. 46, no. 10, pp. 1887–1908, 1998.

[BRA 95] BRAUN J., SAMBRIDGE M., "A numerical method for solving partial differential equations on highly irregular evolving grids", *Nature*, vol. 376, pp. 655–660, 1995.

[BRI 48] BRIDGMAN P.W., "Rough compressions of 177 substances to 40,000 KG/CM2", *Proceedings of the American Academy of Arts and Sciences*, vol. 76, no. 3, pp. 71–87, 1948.

[BRI 53] BRIDGMAN P., ŠIMON I., "Effects of very high pressures on glass", *Journal of Applied Physics*, vol. 24, no. 4, pp. 405–413, 1953.

[BRO 99] BROUGHTON J., ABRAHAM F., BERNSTEIN N. *et al.*, "Concurrent coupling of length scales: methodology and application", *Physical Review B*, vol. 60, no. 4, pp. 2391–2403, 1999.

[BRÜ 70] BRÜCKNER R., "Properties and structure of vitreous silica. I", *Journal of Non-Crystalline Solids*, vol. 5, no. 2, pp. 123–175, 1970.

[BRÜ 71] BRÜCKNER R., "Properties and structure of vitreous silica. II", *Journal of Non-Crystalline Solids*, vol. 5, no. 3, pp. 177–216, 1971.

[CAI 00] CAI W., DEKONING M., BULATOV V. *et al.*, "Minimizing boundary reflections in coupled-domain simulations", *Physical Review Letters*, vol. 85, pp. 3213–3216, 2000.

[CEL 83] CELEP Z., BAZANT Z.P., "Spurious reflection of elastic waves due to gradually changing finite element size", *International Journal for Numerical Methods in Engineering*, vol. 19, pp. 631–646, 1983.

[CHA 10] CHAMOIN L., PRUDHOMME S., BEN DHIA H. *et al.*, "Ghost forces and spurious effects in atomic-to-continuum coupling methods by the Arlequin approach", *International Journal for Numerical Methods in Engineering*, vol. 83, pp. 1081–1113, 2010.

[CHE 01] CHEN J., WU C., YOON S. *et al.*, "A stabilized conforming nodal integration for Galerkin mesh-free methods", *International Journal for Numerical Methods in Engineering*, vol. 50, pp. 435–466, 2001.

[CHI 11] CHINESTA F., CUETO E., CESCOTTO S. *et al.*, *Natural Element Method for the Simulation of Structures and Processes*, ISTE, London and John Wiley & Sons, New York, 2011.

[CHR 62] CHRISTIANSEN E., KISTLER S., GOGARTY W., "Irreversible compressibility of silica glass as a means of determining the distribution of force in high-pressure cells", *Journal of the American Ceramic Society*, vol. 45, no. 4, pp. 172–177, 1962.

[COH 61] COHEN H., ROY R., "Effects of ultra high pressures on glass", *Journal of the American Ceramic Society*, vol. 44, no. 10, pp. 523–524, 1961.

[COH 62] COHEN H., ROY R., Reply to "Comments on 'Effects of Ultrahigh Pressures on Glass'", *Journal of the American Ceramic Society*, vol. 45, no. 8, pp. 398–399, 1962.

[COL 06] COLOMBIER J.P., COMBIS P., ROSENFELD A. *et al.*, "Optimized energy coupling at ultrafast laser-irradiated metal surfaces by tailoring intensity envelopes: consequences for material removal from Al samples", *Physical Review B*, vol. 74, no. 22, pp. 224106–224121, 2006.

[COT 89] COTTET F., BOUSTIE M., "Spallation studies in aluminum targets using shock waves induced by laser irradiation at various pulse durations", *Journal of Applied Physics*, vol. 66, p. 4067, 1989.

[CUE 00] CUETO E., DOBLARÉ M., GRACIA L., "Imposing essential boundary conditions in the natural element method by means of density-scaled -shapes", *International Journal for Numerical Methods in Engineering*, vol. 49, pp. 519–546, 2000.

[CUE 03] CUETO E., SUKUMAR N., CALVO B. ET AL, "Overview and recent advances in natural neighbour Galerkin methods", *Archives of Computational Methods in Engineering*, vol. 10, no. 4, pp. 307–384, 2003.

[CUR 03] CURTIN W., MILLER R., "Atomistic /continuum coupling in computational materials science", *Modeling and Simulation in Materials Science and Engineering*, vol. 11, no. 3, pp. R33–R68, 2003.

[D'AD 01] D'ADDETTA G.A., KUN F., RAMM E. ET AL, "From solids to granulates – discrete element simulations of fracture and fragmentation processes in geomaterials", *Continuous and Discontinuous Modelling of Cohesive-Frictional Materials*, Springer Berlin Heidelberg, of Lecture Notes in Physics, vol. 568, pp. 231–258, 2001.

[DAC 59] DACHILLE F., ROY R., "High pressure studies of the silica isotypes", *Zeitschrift Kristallographie*, vol. 111, pp. 451–461, 1959.

[DAH 08] DAHOTRE N.B., HARIMKAR S.P., "Laser shock processing", *Laser Fabrication and Machining of Materials*, Springer, pp. 477–498, 2008.

[DE 92] DE RESSEGUIER T., Etude expérimentale et numérique du comportement de matériaux de type verre soumis à des ondes de choc, PhD Thesis, University of Poitiers, 1992.

[DEL 34] DELAUNAY B., "Sur la sphère vide. A la mémoire de Georges Voronoï", *Bulletin de l'Académie des Sciences de l'URSS. Classe des sciences mathématiques et na*, vol. 7, no. 6, pp. 793–800, 1934.

[DOB 08] DOBSON M., LUSKIN M., "Analysis of a force-based quasicontinuum approximation", *ESAIM: Mathematical Modelling and Numerical Analysis*, vol. 42, no. 1, pp. 113–139, 2008.

[DOB 10] DOBSON M., LUSKIN M., ORTNER C., "Accuracy of quasicontinuum approximations near instabilities", *Journal of the Mechanics and Physics of Solids*, vol. 58, no. 10, pp. 1741–1757, 2010.

[DOL 99] DOLBOW J., BELYTSCHKO T., "Numerical integration of the Galerkin weak form in meshfree methods", *Computational Mechanics*, vol. 23, pp. 219–230, 1999.

[EDE 01] EDELSBRUNNER H., *Geometry and Topology for Mesh Generation*, Cambridge University Press, 2001.

[FAB 98] FABBRO R., PEYRE P., BERTHE L. *et al.*, "Physics and applications of laser-shock processing", *Journal of Laser Applications*, vol. 10, no. 6, pp. 265–278, 1998.

[FEL 92] FELICI H.M., A coupled Eulerian/Lagrangian method for the solution of three-dimensional vortical flows, PhD Thesis, Massachusetts Institute of Technology, 1992.

[FEY 00] FEYEL F., CHABOCHE J., "Multiscale approach for modelling the elastoviscoplasitic behavior of long fibre SiC/Ti composite materials", *Computer Methods in Applied Mechanics and Engineering*, vol. 183, pp. 309–330, 2000.

[FIS 07] FISH J., NUGGEHALLY M., SHEPHARD M. *et al.*, "Concurrent AtC coupling based on a blend of the continuum stress and the atomistic force", *Computer Methods in Applied Mechanics and Engineering*, vol. 196, pp. 4548–4560, 2007.

[FRO 93] FROMMHOLD L., *Collision-Induced Absorption in Gases*, Cambridge University Press, 1993.

[GON 04] GONZÁLEZ D., CUETO E., MARTÍNEZ M.A. *et al.*, "Numerical integration in natural neighbour Galarkin methods", *International Journal for Numerical Methods in Engineering*, vol. 60, pp. 2077–2114, 2004.

[GRA 78] GRAHAM R.A., ASAY J.R., "Measurements of wave profiles in shock-loaded solids", *High Temperatures-High Pressures*, vol. 10, pp. 355–390, 1978.

[GRI 21] GRIFFITH A., "The phenomena of rupture and flow in solids", *Philosophical Transactions of the Royal Society of London. Series A*, vol. 221, pp. 163–198, 1921.

[GRO 12] GROSS T.M., "Deformation and cracking behavior of glasses indented with diamond tips of various sharpness ", *Journal of Non-Crystalline Solids*, vol. 358, no. 24, pp. 3445–3452, 2012.

[GUI 07] GUIDAULT P.A., BELYTSCHKO T., "On the L2 and the H1 couplings for an overlapping domain decomposition method using Lagrange multipliers", *International Journal for Numerical Methods in Engineering*, vol. 70, pp. 322–350, 2007.

[GUI 09] GUIDAULT P.A., BELYTSCHKO T., "Bridging domain methods for coupled atomistic-continuum models with L2 or H1 couplings", *International Journal for Numerical Methods in Engineering*, vol. 77, pp. 1566–1592, 2009.

[HAL 94] HALL W.S., *Boundary Element Method*, Springer, vol. 27, 1994.

[HAN 11] HAN I., LI Y., "Chapter Interaction between pulsed laser and materials", in JAKUBCZAK K. (ed.), *Lasers Applications in Science and Industry*, InTech, 2011.

[HIY 02] HIYOSHI H., SUGIHARA K., "Improving continuity of Voronoi-based interpolation over Delaunay spheres", *Computational Geometry*, vol. 22, pp. 167–183, 2002.

[HOL 93] HOLMQUIST T.J., JOHNSON G.R., "A computational constitutive model for concrete subjected to large strains, high strain rate and high pressure", *14th International Symposium on Ballistics, Quebec, Canada*, pp. 591–600, 1993.

[IOR 08] IORDANOFF I., BATTENTIER A., NEAUPORT J. *et al.*, "A discrete element model to investigate sub-surface damage due to surface polishing", *Tribology International*, vol. 41, no. 11, pp. 957–964, 2008.

[IYE 11] IYER M., GAVINI V., "A field theoretical approach to the quasi-continuum method", *Journal of the Mechanics and Physics of Solids*, vol. 59, no. 8, pp. 1506–1535, 2011.

[JEB 13a] JEBAHI M., Discrete-continuum coupling method for simulation of laser-induced damage in silica glass, PhD Thesis, Bordeaux 1 University, 2013.

[JEB 13b] JEBAHI M., ANDRÉ D., DAU F. *et al.*, "Simulation of Vickers indentation of silica glass", *Journal of Non-Crystalline Solids*, vol. 378, pp. 15–24, 2013.

[JEB 13c] JEBAHI M., CHARLES J., DAU F. *et al.*, "3D coupling approach between discrete and continuum models for dynamic simulations (DEM-CNEM)", *Computer Methods in Applied Mechanics and Engineering*, vol. 255, pp. 196–209, 2013.

[JEB 14] JEBAHI M., DAU F., CHARLES J.L. *et al.*, "Multiscale modeling of complex dynamic problems: an overview and recent developments", *Archives of Computational Methods in Engineering*, Springer, Netherlands, pp. 1–38, 2014.

[JEB 15] JEBAHI M., ANDRÉ D., TERREROS I. *et al.*, *Discrete Element Method to Model 3D Continuous Materials*, ISTE, London and John Wiley & Sons, New York, 2015.

[JI 07] JI H., Mécanique et physique de l'indentation du verre , PhD Thesis, University of Rennes 1, 2007.

[JOH 94] JOHNSON G.R., HOLMQUIST T.J., "An improved computational constitutive model for brittle materials", *High Pressure Science and Technology, American Institute Physics*, pp. 981–984, 1994.

[KAL 97] KALJEVIC I., SAIGAL S., "An improved element free Galerkin formulation", *International Journal for Numerical Methods in Engineering*, vol. 40, pp. 2953–2974, 1997.

[KAN 98] KANEL G., "Some new data on deformation and fracture of solids under shock-wave loading", *Journal of the Mechanics and Physics of Solids*, vol. 35, no. 10, pp. 1869–1886, 1998.

[KER 08] KERYVIN V., Contribution à l'étude des mécanismes de déformation et de fissuration des verres, thesis, University of Rennes 1, 2008.

[KHA 05] KHAN S.A., "Wavelength-dependent modifications in Helmholtz optics", *International Journal of Theoretical Physics*, vol. 44, no. 1, pp. 95–125, 2005.

[KON 81] KONDO K., IIO S., SAWAOKA A., "Nonlinear pressure dependence of the elastic moduli of fused quartz up to 3 GPa", *Journal of Applied Physics*, vol. 52, no. 4, pp. 2826–2831, 1981.

[LAL 91] LALLE P., Utilisation de l'eau, de la silice fondue et de deux verres de la société Shott comme fenêtre, Report, DAM/CESTA/ESD DO 1533, 1991.

[LAW 93] LAWN B., *Fracture of Brittle Solids – second edition*, Cambridge University Press, 1993.

[LAW 03] LAWN B.R., "Fracture and deformation in brittle solids: a perspective on the issue of scale", *Journal of Material Research*, vol. 19, pp. 22–29, 2003.

[LE 03] LE HOUÉROU V., SANGLEBOEUF J.C., DÉRIANO S. *et al.*, "Surface damage of soda-lime-silica glasses: indentation scratch behavior", *Journal of Non-Crystalline Solids*, vol. 316, pp. 54–63, 2003.

[LEE 03] LEE C.K., ZHOU C.E., "On error estimation and adaptive refinement for element free galerkin method: part I: stress recovery and a posteriori error estimation", *Computer and Structures*, vol. 82, nos. 4–5, pp. 4293–443, 2003.

[LES 06] LESCOUTE E., Étude de la fragmentation dynamique de métaux sous choc laser, PhD Thesis, Ecole Nationale Supérieure de Mécanique et d'Aérotechnique, 2006.

[LI 02] LI S., LIU W.K., "Meshfree and particle methods and their applications", *Applied Mechanics Reviews*, vol. 55, no. 1, pp. 1–34, 2002.

[LI 06] LI S., LIU X., AGRAWAL A. *et al.*, "Perfectly matched multiscale simulations for discrete lattice systems: extension to multiple dimensions", *Physical Review B*, vol. 74, p. 045418, 2006.

[LI 14] LI X., MING P., "On the effect of ghost force in the quasicontinuum method: dynamic problems in one dimension", *Communications in Computational Physics*, vol. 15, pp. 647–676, 2014.

[LIS 80] LISZKA T., ORKISZ J., "The finite difference method at arbitrary irregular grids and its applications in applied mechanics", *Computers and Structures*, vol. 11, pp. 83–95, 1980.

[LIU 01] LIU G.R., GU Y.T., "A point interpolation method for two-dimensional solids", *International Journal for Numerical Methods in Engineering*, vol. 50, pp. 937–951, 2001.

[LIU 02] LIU M.B., LIU G.R., LAM K.Y., "Coupling meshfree particle method with molecular dynamics – a novel approach for multi-scale simulations ", *Proceedings of the 1st Asian Workshop on Meshfree Methods*, Advances in Meshfree and X-FEM Methods, pp. 211–216, 2002.

[LIU 03] LIU G.R., LIU M.B., *Smoothed Particle Hydrodynamics: A Meshfree Particle Method*, World Scientific Publishing Co. Pte. Ltd, 2003.

[LIU 10] LIU M.B., LIU G., "Smoothed particle hydrodynamics (SPH): an overview and recent developments", *Archives of Computational Methods in Engineering*, vol. 17, pp. 25–76, 2010.

[LU 05] LU G., KAXIRAS E., "Overview of multiscale simulations of materials", in RIETH M., SCHOMMERS W. (eds), *Handbook of Theoretical and Computational Nanotechnology*, American Scientific Publishers, 2005.

[LUB 08] LUBLINER J., *Plasticity Theory*, Dover Publications, 2008.

[LUC 77] LUCY L.B., "Numerical approach to testing the fission hypothesis", *Astronomical Journal*, vol. 82, pp. 1013–1024, 1977.

[MAC 63] MACKENZIE J., "High-pressure effects on oxide glasses: I, densification in rigid state", *Journal of the American Ceramic Society*, vol. 46, no. 10, pp. 461–470, 1963.

[MAG 98] MAGNIER S.A., DONZÉ F.V., "Numerical simulations of impacts using a discrete element method", *Mechanics of Cohesive-Frictional Materials*, John Wiley & Sons, Ltd., vol. 3, no. 3, pp. 257–276, 1998.

[MAI 95] MAIR H.U., "Hydrocode methodologies for underwater explosion structure medium/interaction", *Shock and Vibration*, vol. 2, pp. 227–248, 1995.

[MAI 08] MAIRE P.H., BREIL J., GALERA S., "A cell-centred arbitrary Lagrangian–Eulerian (ALE) method", *International Journal for Numerical Methods in Fluids*, vol. 56, pp. 1161–1166, 2008.

[MAN 96] MANDELL D.A., WINGATE C.A., SCHWALBE L.A., "Simulation of a ceramic impact experiment using the SPHINX SPH code", *16th International Symposium on Ballistics, San Francisco, CA*, 1996.

[MAR 80] MARSH S.P., *LASL Shock Hugoniot Data*, University of California Press, Berkeley, 1980.

[MEL 96] MELENKA J.M., BABUKA I., "The partition of unity finite element method: basic theory and applications", *Computer Methods in Applied Mechanics and Engineering*, vol. 139, nos. 1–4, pp. 289–314, 1996.

[MEY 81] MEYERS M.A., MURR L.E., *Shock Waves and High-Strain-Rate Phenomena in Metals: Concepts and Applications* , Plenum, 1981.

[MEY 88] MEYER A., "An efficient implementation of LU decomposition in C", *Advances in Engineering Software*, vol. 10, no. 3, pp. 123–130, 1988.

[MEY 94] MEYERS M.A., *Dynamic Behavior of Materials*, John Wiley & Sons, 1994.

[MIC 83] MICHALSKE T., FREIMAN S., "A molecular mechanism for stress corrosion in vitreous silica", *Journal of the American Ceramic Society*, vol. 66, no. 4, pp. 284–288, 1983.

[MIC 07] MICHEL Y., Phénomène d'impact à haute vitesse sur cibles minces fragiles: application au projet de laser mégajoule et à la problématique des débris spatiaux, PhD Thesis, University of Toulouse III, 2007.

[MOË 99] MOËS N., DOLBOW J., BELYTSCHKO T., "A finite element method for crack growth without remeshing", *International Journal for Numerical Methods in Engineering*, vol. 46, pp. 131–150, 1999.

[NUG 07] NUGGEHALLY M.A., SHEPHARD M.S., PICU C.R. *et al.*, "Adaptive model selection procedure for concurrent multiscale problems", *International Journal for Multiscale Computational Engineering*, vol. 5, no. 5, pp. 369–386, 2007.

[OÑA 96] OÑATE E., IDELSOHN S., ZIENKIEWICZ O.C. *et al.*, "A stabilized finite point method for analysis of fluid mechanics problems", *Computer Methods in Applied Mechanics and Engineering*, vol. 139, pp. 315–346, 1996.

[OÑA 98] OÑATE E., IDELSOHN S., "A mesh-free finite point method for advective-diffusive transport and fluid flow problems", *Computational Mechanics*, vol. 21, pp. 283–292, 1998.

[OÑA 00] OÑATE E., SACCO C., IDELSOHN S., "A finite point method for incompressible flow problems", *Computing and Visualization in Science*, vol. 3, pp. 67–75, 2000.

[ORO 49] OROWAN E., "Fracture and strength of solids", *Reports on Progress in Physics*, vol. 12, p. 185, 1949.

[ORT 01] ORTIZ M., KNAP J., "An analysis of the quasicontinuum method", *Journal of the Mechanics and Physics of Solids*, vol. 49, no. 9, pp. 1899–1923, 2001.

[PAR 05a] PARK H., KARPOV E., LIU W. *et al.*, "The bridging scale for two-dimensional atomistic/continuum coupling", *Philosophical Magazine*, vol. 85, no. 1, pp. 79–113, 2005.

[PAR 05b] PARKA H., KARPOVB E., KLEINA P. *et al.*, "Three-dimensional bridging scale analysis of dynamic fracture", *Journal of Computational Physics*, vol. 207, no. 2, pp. 588–609, 2005.

[PHI 02] PHIPPS C.R., LUKE J.R., MCDUFF G.G. *et al.*, "Laser ablation powered mini-thruster", *High-Power Laser Ablation IV*, vol. 4760, pp. 833–842, 2002.

[POL 93] POLIAN A., GRIMSDITCH M., "Sound velocities and refractive index of a-SiO2 to 25 GPa", *Physical Review B*, vol. 47, p. 13979, 1993.

[RAN 70] RANKINE W.J.M., "On the thermodynamic theory of waves of finite longitudinal disturbance", *Philosophical Transactions of the Royal Society of London*, vol. 160, pp. 277–288, 1870.

[RAN 87] RANKINE P.H., "Sur la propagation du mouvement dans les corps et spécialement dans les gaz parfaits, 1e Partie", *Journal de l'École Polytechnique. Paris.*, vol. 57, pp. 3–97, 1887.

[RAN 89] RANKINE P.H., "Sur la propagation du mouvement dans les corps et spécialement dans les gaz parfaits, 2e Partie", *Journal de l'École Polytechnique. Paris.*, vol. 58, pp. 1–125, 1889.

[RAN 96] RANDLES P.W., LIBERSKY L.D., "Smoothed particle hydrodynamics: some recent improvements and applications", *Computer Methods in Applied Mechanics and Engineering*, vol. 139, pp. 375–408, 1996.

[RAT 03] RATEAU G., Méthode Arlequin pour les problèmes mécaniques multi-échelles Applications à des problèmes de jonction et de fissuration de structures élancées, PhD Thesis, École Centrale Paris, 2003.

[RIC 07] RICHARD D., IORDANOFF I., BERTHIER Y. *et al.*, "Friction coefficient as a macroscopic view of local dissipation", *Journal of Tribology*, ASME, vol. 129, no. 4, pp. 829–835, 2007.

[RIN 52] RINEHART J.S., "Scabbing of metals under explosive attack: multiple scabbing ", *Journal of Applied Physics*, vol. 23, no. 11, pp. 1229–1233, 1952.

[RIO 96] RIOU P., Contribution à l'étude de l'endommagement du carbure de Silicium lors d'un impact de basse énergie: application aux blindages, PhD Thesis, Ecole Nationale des Mines de Paris, 1996.

[ROD 03] RODNEY D., "Mixed atomistic/continuum methods: static and dynamic quasicontinuum methods", *Proceedings of the NATO Conference: Thermodynamics, Microstructures and Plasticity*, pp. 265–274, 2003.

[ROU 00] ROUXEL T., SANGLEBOEUF J.C., "The brittle to ductile transition in a soda-lime-silica glass", *Journal of Non-Crystalline Solids*, vol. 271, no. 3, pp. 224–235, 2000.

[ROU 04] ROUGIER E., MUNJIZA A., JOHN N., "Numerical comparison of some explicit time integration schemes used in DEM, FEM/DEM and molecular dynamics", *International Journal for Numerical Methods in Engineering*, vol. 62, pp. 856–879, 2004.

[ROU 08] ROUXEL T., JI H., HAMMOUDA T. *et al.*, "Poisson's ratio and the densification of glass under high pressure", *Physical Review letters*, vol. 100, no. 22, p. 225501, 2008.

[ROU 10] ROUXEL T., JI H., AUGEREAU F. *et al.*, "Indentation deformation mechanism in glass: densification versus shear flow", *Journal of Applied Physics*, vol. 107, 2010.

[SHE 98] SHEWCHUCK J.R., "Tetrahedral mesh generation by Delaunay refinement", *Proceedings of the 14th Annual Symposium on Computational Geometry, Minneapolis, Minnesota, Association for Computing Machinery*, pp. 86–95, June 1998.

[SHE 99] SHENOY V., Quasicontinuum models of atomic-scale mechanics, PhD Thesis, Brown University, 1999.

[SHI 04] SHIMOKAWA T., MORTENSEN J., SCHIØTZ J. *et al.*, "Matching conditions in the quasicontinuum method: removal of the error introduced at the interface between the coarse-grained and fully atomistic region", *Physical Review B*, vol. 69, no. 21, p. 214104, 2004.

[SHI 09] SHIU W., DONZÉ F.V., DAUDEVILLE L., "Discrete element modelling of missile impacts on a reinforced concrete target", *International Journal of Computer Applications in Technology*, vol. 34, no. 1, pp. 33–41, 2009.

[SIB 80] SIBSON R., "A vector identity for the Dirichlet tessellations", *Mathematical Proceedings of the Cambridge Philosphical Society*, vol. 87, pp. 151–155, 1980.

[SIB 81] SIBSON R., "A brief description of natural neighbour interpolation", in BARNETT V. (ed.), *Interpreting Multivariate Data,* John Wiley, Chichester, pp. 21–36, 1981.

[SMI 85] SMITH G.D., *Numerical Solution of Partial Differential Equations: Finite Difference Methods*, 3rd ed., Oxford University Press, 1985.

[SMI 99] SMIRNOVA J.A., ZHIGILEI L.V., GARRISON B.J., "A combined molecular dynamics and finite element method technique applied to laser induced pressure wave propagation", *Computer Physics Communications*, vol. 118, pp. 11–16, 1999.

[STE 04] STEIN E., DE BORST R., HUGHES T.J.R., *Encyclopedia of Computational Mechanics*, John Wiley & Sons Ltd., vol. 1, 2004.

[SUG 81] SUGIURA H., KONDO K., SAWAOKA A., "Dynamic response of fused quartz in the permanent densification region", *Journal of Applied Physics*, vol. 52, no. 5, pp. 3375–3382, 1981.

[SUG 96] SUGIURA H., IKEDA R., KONDO K. *et al.*, "Densified silica glass after shock compression", *Journal of Applied Physics*, vol. 81, pp. 1651–1655, 1996.

[SUK 98] SUKUMAR N., MORAN B., BELYTSCHKO T., "The natural element method in solid mechanics", *International Journal for Numerical Methods in Engineering*, vol. 43, pp. 839–887, 1998.

[TAD 96a] TADMOR E.B., ORTIZ M., PHILLIPS R., "Quasicontinuum analysis of defects in solids", *Philosophical Magazine A*, vol. 73, no. 6, pp. 1529–1563, 1996.

[TAD 96b] TADMOR E., PHILLIPS R., ORTIZ M., "Mixed atomistic and continuum models of deformation in solids", *Langmuir*, vol. 12, no. 19, pp. 4529–4534, 1996.

[TAN 09] TAN Y., YANG D., SHENG Y., "Discrete element method (DEM) modelling of fracture and damage in the machining process of polycrystalline sic", *Journal of the European Ceramic Society*, vol. 29, no. 6, pp. 1029–1037, 2009.

[TAN 11] TANI G., ORAZI L., FORTUNATO A. *et al.*, "Warm laser shock peening: new developments and process optimization", *Manufacturing Technology*, vol. 60, pp. 219–222, 2011.

[TER 13] TERREROS I., IORDANOFF I., CHARLES J., "Simulation of continuum heat conduction using DEM domains", *Computational Materials Science*, vol. 69, no. 0, pp. 46–52, 2013.

[TO 05] TO A., LI S., "Perfectly matched multiscale simulations", *Physical Review B*, vol. 72, p. 035414, 2005.

[TRA 94] TRAVERSONI L., "Natural neighbour finite elements", *International Conference on Hydraulic Engineering Software Hydrosoft Proceedings*, vol. 2, pp. 291–297, 1994.

[VER 07] VERSTEEG H.K., MALALASEKERA W., *An Introduction to Computational Fluid Dynamics: The Finite Volume Method*, 2nd ed., Pearson Education Limited, 2007.

[VOR 07] VORONOI G., "Nouvelles applications des paramètres continus à la théorie des formes quadratiques", *Journal für die Reine und Angewandte Mathematik*, vol. 133, pp. 97–178, 1907.

[WAC 62] WACKERLE J., "Shock wave compression of quartz", *Journal of Applied Physics*, vol. 33, no. 3, pp. 922–937, 1962.

[WAG 03] WAGNER G., LIU W., "Coupling of atomistic and continuum simulations using a bridging scale decomposition", *Journal of Computational Physics*, vol. 190, no. 1, pp. 249–274, 2003.

[WAL 55] WALSH J.M., CHRISTIAN R.H., "Equation of state of metals from shock wave measurements", *Physical Review*, vol. 97, pp. 1544–1556, 1955.

[WEI 62] WEIR C.E., SPINNER S., Comments on "Effects of Ultrahigh Pressures on Glass", *Journal of the American Ceramic Society*, vol. 45, no. 4, p. 196, 1962.

[WHI 63] WHITE R.M., "Elastic wave generation by electron bombardment or electromagnetic wave absorption", *Journal of Applied Physics*, vol. 34, no. 7, pp. 2123–2124, 1963.

[WIE 67] WIEDERHORN S.M., "Influence of water vapor on crack propagation in soda-lime glass", *Journal of the American Ceramic Society*, vol. 50, no. 8, pp. 407–414, 1967.

[WEI 02] WEINAN E., HUANGA Z., "A dynamic atomistic-continuum method for the simulation of crystalline materials", *Journal of Computational Physics*, vol. 182, pp. 234–261, 2002.

[WIE 04] WIEDERHORN S.M., GUIN J.P., "Fracture of silicate glasses: ductile or brittle?", *Physical Review letters*, vol. 92, no. 21, p. 215502, 2004.

[WEI 06] WEINAN E., LU J., YANG J., "Uniform accuracy of the quasicontinuum method", *Physical Review B*, vol. 74, no. 21, p. 214115, 2006.

[WOO 03] WOOD R.M., *Laser-Induced Damage of Optical Materials*, Institute of Physics Publishing, 2003.

[WU 05] WU H.C., *Continuum Mechanics and Plasticity*, Chapman & Hall, 2005.

[XIA 04] XIAO S.P., BELYTSCHKO T., "A bridging domain method for coupling continua with molecular dynamics", *Computer Methods in Applied Mechanics and Engineering*, vol. 193, nos. 17–20, pp. 1645–1669, 2004.

[XIE 09] XIE W., LIU Z., YOUNG Y.L., "Application of a coupled Eulerian-Lagrangian method to simulate interactions between deformable composite structures and compressible multiphase flow", *International Journal for Numerical Methods in Engineering*, vol. 80, pp. 1497–1519, 2009.

[XU 09] XU M., GRACIE R., BELYTSCHKO T., "Multiscale modeling with extended bridging domain method", in FISH J. (ed.), *Bridging the Scales in Science and Engineering*, Oxford Press, 2009.

[YOS 05] YOSHIDA S., SANGLEBOEUF J., ROUXEL T., "Quantitative evaluation of indentation-induced densification in glass", *Journal of Materials Research*, vol. 20, no. 12, pp. 3404–3412, 2005.

[YVO 04a] YVONNET J., Nouvelles approches sans maillage basées sur la méthode des éléments naturels pour la simulation numérique des procédés de mise en forme, PhD Thesis, École Nationale Supérieure d'Arts et Métiers, 2004.

[YVO 04b] YVONNET J., RYCKELYNCK D., LORONG P. *et al.*, "A new extension of the natural element method for non-convex and discontinuous problems: the constrained natural element method (C-NEM)", *International Journal for Numerical Methods in Engineering*, vol. 60, pp. 1451–1474, 2004.

[YVO 05] YVONNET J., CHINESTA F., LORONG P. *et al.*, "The constrained natural element method (C-NEM) for treating thermal models involving moving interfaces", *International Journal of Thermal Sciences*, vol. 44, pp. 559–569, 2005.

[ZHA 94] ZHA C.S., HEMLEY R.J., HO-KWANG M. *et al.*, "Acoustic velocities and refractive index of SiO_2 glass to 57.5 GPa by Brillouin scattering", *Physical Review B*, vol. 50, no. 18, pp. 13105–13112, 1994.

[ZHU 98] ZHU T., ALTRUNI S.N., "A modified collocation method and a penalty formulation for enforcing the essential boundary conditions in the element free Galerkin methods", *International Journal for Numerical Methods in Engineering*, vol. 21, pp. 211–222, 1998.

[ZHU 01] ZHU Z., LU X., LI J., "A study of domain decomposition and parallel computation", *Acta Mechanica*, vol. 150, nos. 3–4, pp. 219–235, 2001.

[ZIE 05a] ZIENKIEWICZ O.C., TAYLOR R.L., *The Finite Element Method for Solid and Structural Mechanics*, Elsevier, 2005.

[ZIE 05b] ZIENKIEWICZ O.C., TAYLOR R.L., NITHIARASU P., *The Finite Element Method for Fluid Dynamics*, Elsevier, 2005.

[ZIE 05c] ZIENKIEWICZ O.C., TAYLOR R.L., ZHU J.Z., *Finite Element Method: Its Basis & Fundamentals*, Elsevier, 2005.

Index

Other titles from

iSTE

in

Numerical Methods in Engineering

2015

JEBAHI Mohamed, ANDRÉ Damien, TERREROS Inigo, IORDANOFF Ivan
Discrete Element Method to Model 3D Continuous Materials

2014

BECKERS Benoit, BECKERS Pierre
Reconciliation of Geometry and Perception in Radiation Physics

BERGHEAU Jean-Michel
Thermo-Mechanical Industrial Processes: Modeling and Numerical Simulation

BONNEAU Dominique, FATU Aurelian, SOUCHET Dominique
4-volume series
Hydrodynamic Bearings – Volume 1
Mixed Lubrication in Hydrodynamic Bearings – Volume 2
Thermo-hydrodynamic Lubrication in Hydrodynamic Bearings – Volume 3
Internal Combustion Engine Bearings Lubrication in Hydrodynamic Bearings – Volume 4

DESCAMPS Benoît
Physical Form Finding of Lightweight Structures

2013

YASTREBOV Vladislav A.
Numerical Methods in Contact Mechanics

2012

DHATT Gouri, LEFRANÇOIS Emmanuel, TOUZOT Gilbert
Finite Element Method

2011

CHINESTA Francisco *et al.*
Natural Element Method for the Simulation of Structures and Processes

DAVIM Paulo J.
Finite Element Method in Manufacturing Processes

POMMIER Sylvie, GRAVOUIL Anthony, MOËS Nicolas, COMBESCURE Alain
Extended Finite Element Method for Crack Propagation

2010

SOUZA DE CURSI Eduardo, SAMPAIO Rubens
Modeling and Convexity

2008

BERGHEAU Jean-Michel, FORTUNIER Roland
Finite Element Simulation of Heat Transfer

EYMARD Robert
Finite Volumes for Complex Applications V

FREY Pascal, GEORGE Paul-Louis
Mesh Generation – 2nd edition

GAY Daniel, GAMBELIN Jacques
Modeling and Dimensioning of Structures

2005

BENKHALDOUN Fayssal
Finite Volumes for Complex Applications IV

Printed and bound by CPI Group (UK) Ltd, Croydon, CR0 4YY

13/09/2026

14961016-0002